COST Action FA0906

Beyond the visible

A handbook of best practice in plant UV photobiology

COST action F0906 'UV4growth'

Beyond the visible
A handbook of best practice in plant UV photobiology

Edited by
Pedro J. Aphalo
with

Andreas Albert   Lars Olof Björn   Andy McLeod   T. Matthew Robson   Eva Rosenqvist

EUROPEAN COOPERATION
IN SCIENCE AND TECHNOLOGY

ESF provides the COST office through an EC contract

COST is supported by the EU RTD framework programme

Full citation:

Aphalo, P. J.; Albert, A.; Björn, L. O.; McLeod, A.; Robson, T. M.; Rosenqvist, E. (eds.) 2012. *Beyond the visible: A handbook of best practice in plant UV photobiology.* COST Action FA0906 *UV4growth*. Helsinki: University of Helsinki, Division of Plant Biology. ISBN 978-952-10-8362-4 (Paperback), 978-952-10-8363-1 (PDF). xxx + 176 pp.

First edition, first printing: November 2012.
First edition, second corrected printing: February 2013.

UNIVERSITY OF HELSINKI
FACULTY OF BIOLOGICAL AND ENVIRONMENTAL SCIENCES

Published by the University of Helsinki, Department of Biosciences, Division of Plant Biology, Helsinki, Finland.
Paperback edition printed through CreateSpace Independent Publishing Platform
ISBN-13: 978-952-10-8362-4 (Paperback)
ISBN-13: 978-952-10-8363-1 (PDF)

Typeset with LaTeX in Lucida Bright and Lucida Sans using the KOMA-Script book class.
Most data plots were produced with the R System for statistics with package ggplot2. Flowcharts drawn with yEd.
Edited with WinEdt.

COST

COST —the acronym for European Cooperation in Science and Technology— is the oldest and widest European intergovernmental network for cooperation in research. Established by the Ministerial Conference in November 1971, COST is presently used by the scientific communities of 36 European countries to cooperate in common research projects supported by national funds.

The funds provided by COST —less than 1% of the total value of the projects— support the COST cooperation networks (COST Actions) through which, with EUR 30 million per year, more than 30 000 European scientists are involved in research having a total value which exceeds EUR 2 billion per year. This is the financial worth of the European added value which COST achieves.

A "bottom up approach" (the initiative of launching a COST Action comes from the European scientists themselves), "à la carte participation" (only countries interested in the Action participate), "equality of access" (participation is open also to the scientific communities of countries not belonging to the European Union) and "flexible structure" (easy implementation and light management of the research initiatives) are the main characteristics of COST.

As precursor of advanced multidisciplinary research COST has a very important role for the realisation of the European Research Area (ERA) anticipating and complementing the activities of the Framework Programmes, constituting a "bridge" towards the scientific communities of emerging countries, increasing the mobility of researchers across Europe and fostering the establishment of "Networks of Excellence" in many key scientific domains such as: Biomedicine and Molecular Biosciences; Food and Agriculture; Forests, their Products and Services; Materials, Physical and Nanosciences; Chemistry and Molecular Sciences and Technologies; Earth System Science and Environmental Management; Information and Communication Technologies; Transport and Urban Development; Individuals, Societies, Cultures and Health. It covers basic and more applied research and also addresses issues of pre-normative nature or of societal importance.

Web: http://www.cost.eu

Foreword

The discovery of the thinning of the stratospheric ozone layer in the 1980s, and subsequent measurements of increased penetration of UV-B in the biosphere, triggered extensive research in the photobiology of UV radiation. Initially, the central aim of research was to assess the impacts of increases in UV-B radiation on various organisms and ecosystems. More recently, the question how current levels of UV affect life-processes has become increasingly important. Several decades of UV research have now led to a much more detailed understanding of the, rather broad, impacts of UV on plant, microbial and animal life. Emphasis has gradually shifted away from a stress-dominated view (sunburn in humans and farm animals; macroscopic damage and growth inhibition in plants) to a more balanced vision which also includes many subtle, regulatory UV-effects. For example, many vertebrates gather information from UV wavelengths which they can perceive due to their tetrachromatic colour vision. Birds appear to use UV wavelengths for orientation and navigation, while both birds and lizards have UV-reflective features that play a role in mate choice. Several rodents, including house mice, can perceive UV-light, while their urine fluoresces under UV, consequently it has been suggested that these rodents may use UV cues for intraspecific signalling. Another mammal able to perceive UV wavelengths is the reindeer, which is thought to obtain information from differential UV reflections of Arctic vegetation. Plants are not to be left out of this wonderful (for human's invisible) UV world: One of the most important discoveries in plant UV-B biology has been the identification of a specific UV-B photoreceptor. Thus, plants also "see" UV-B and the so-called UVR8 photoreceptor has, among others, been implicated in controlling the development of plant morphology in UV-B exposed plants.

Notwithstanding these fascinating advances in UV biology, the development of an overarching vision on the biological role of UV wavelengths remains elusive. This is partly due to the use of a rather diverse range of UV-exposure and quantification technologies, in combination with action spectra that are not always appropriate. Consequent variations in applied dose and spectrum have affected the reproducibility of research across different laboratories, and have sometimes necessitated extensive trialling to repeat published data. A related issue concerns the extrapolation of laboratory data, generated indoors using artificial UV sources and/or filtration set-ups, to more ecologically relevant scenarios. There is no doubt that indoor experimentation, under rather unnatural conditions, has generated conceptually critical information about UV-perception, and UV-mediated signalling and gene transcription. Nevertheless, one important lesson learnt from three decades of plant UV-research is that time spent on devising an experimental set-up that is as "environmentally sound" as feasible, is time well spent (climate change biologists, please take note!).

This book entitled "Beyond the visible: A handbook of best practice in plant UV photobiology" is an important contribution towards such sound experimental design, promoting both "good practice" in UV-B manipulation, as well as "standardisation" of methodologies. Writing an authoritative book that will steer experimental approaches over the coming years, can not easily be done by an individual, but rather requires the concerted effort of a team of expert scientists. I commend the main author, Dr. Pedro J. Aphalo, who assembled a team of leading UV-scientists, and I congratulate all the authors on a text that is both accessible as well as in-depth. I also gratefully acknowledge the financial support of COST (European Cooperation in Science and Technology), who through COST Action UV4Growth (FA0906) made it possible for the main authors to meet, coordinate and write. This is surely an excellent example of a concerted, European-wide activity that will boost the plant UV-B research field in Europe and beyond, for years to come.

Happy reading,

Cork, August 2012
Dr. Marcel A. K. Jansen
Chair, COST Action UV4Growth

Contents

Contents

List of Tables

List of Figures

List of Text Boxes

Contributors

Andreas Albert
Research Unit Environmental Simulation (EUS), Helmholtz Zentrum München, 85764 Neuherberg, Germany.

Pedro J. Aphalo
Department of Biosciences, P.O. Box 65, 00014 University of Helsinki, Finland.
mailto:pedro.aphalo@helsinki.fi

Lars Olof Björn
School of Life Science, South China Normal University, Guangzhou 510631, China
and
Lund University, Department of Biology, SE-223 62 Lund, Sweden.

Iván Gómez Ocampo
Instituto de Ciencias Marinas y Limnológicas, Facultad de Ciencias, Universidad Austral de Chile, Valdivia, Chile.

Daniele Grifoni
Institute of Biometeorology (IBIMET), National Research Council (CNR), Florence, Italy.

Anu Heikkilä
Finnish Meteorological Institute, Helsinki, Finland.

Pirjo Huovinen
Instituto de Ciencias Marinas y Limnológicas, Facultad de Ciencias, Universidad Austral de Chile, Valdivia, Chile.

Harri Högmander
Department of Mathematics and Statistics, Faculty of Mathematics and Science, University of Jyväskylä, Finland.

Predrag Kolarž
Laboratory for Atomic Collision Processes, Institute of Physics, 11080 Belgrade, Serbia.

Anders V. Lindfors
Kuopio Unit, Finnish Meteorological Institute, Finland.

Félix López Figueroa
Department of Ecology, University of Málaga, Málaga, Spain.

Andy McLeod
School of GeoSciences, University of Edinburgh, Edinburgh EH9 3JN, Scotland, United Kingdom.

T. Matthew Robson
Department of Biosciences, P.O. Box 65, 00014 University of Helsinki, Finland.

Eva Rosenqvist
Department of Agriculture and Ecology/Crop Science, University of Copenhagen, Denmark.

Åke Strid
Department of Science and Technology, Örebro Life Science Center, Örebro University, Sweden.

Lasse Ylianttila
Radiation and Nuclear Safety Authority Finland, PL 14, 00881 Helsinki, Finland.

Gaetano Zipoli
Institute of Biometeorology (IBIMET), National Research Council (CNR), Florence, Italy.

List of abbreviations and symbols

For quantities and units used in photobiology we follow, as much as possible, the recommendations of the Commission Internationale de l'Éclairage as described by Sliney (2007).

Symbol	Definition
α	absorptance (%).
Δe	water vapour pressure difference (Pa).
ϵ	emittance ($W\,m^{-2}$).
λ	wavelength (nm).
θ	solar zenith angle (degrees).
ν	frequency (Hz or s^{-1}).
ρ	reflectance (%).
σ	Stefan-Boltzmann constant.
τ	transmittance (%).
χ	water vapour content in the air ($g\,m^{-3}$).
A	absorbance (absorbance units).
ANCOVA	analysis of covariance.
ANOVA	analysis of variance.
BSWF	biological spectral weighting function.
c	speed of light in a vacuum.
CCD	charge coupled device, a type of light detector.
CDOM	coloured dissolved organic matter.
CFC	chlorofluorocarbons.
c.i.	confidence interval.
CIE	Commission Internationale de l'Éclairage (International Commission on Illumination); or when refering to an action spectrum, the erythemal action spectrum standardized by CIE.
CTC	closed-top chamber.
DAD	diode array detector, a type of light detector based on photodiodes.
DBP	dibutylphthalate.
DC	direct current.
DIBP	diisobutylphthalate.
DNA(N)	UV action spectrum for 'naked' DNA.
DNA(P)	UV action spectrum for DNA in plants.
DOM	dissolved organic matter.
DU	Dobson units.
e	water vapour partial pressure (Pa).
E	(energy) irradiance ($W\,m^{-2}$).
$E(\lambda)$	spectral (energy) irradiance ($W\,m^{-2}\,nm^{-1}$).
E_0	fluence rate, also called scalar irradiance ($W\,m^{-2}$).
ESR	early stage researcher.
FACE	free air carbon-dioxide enhancement.
FEL	a certain type of 1000 W incandescent lamp.
FLAV	UV action spectrum for accumulation of flavonoids.
FWHM	full-width half-maximum.
GAW	Global Atmosphere Watch.
GEN	generalized plant action spectrum, also abreviated as GPAS (Caldwell, 1971).
GEN(G)	mathematical formulation of GEN by Green et al. (1974) .
GEN(T)	mathematical formulation of GEN by Thimijan et al. (1978).

h	Planck's constant.
h'	Planck's constant per mole of photons.
H	exposure, frequently called dose by biologists ($kJ\,m^{-2}\,d^{-1}$).
H^{BE}	biologically effective (energy) exposure ($kJ\,m^{-2}\,d^{-1}$).
H_p^{BE}	biologically effective photon exposure ($mol\,m^{-2}\,d^{-1}$).
HPS	high pressure sodium, a type of discharge lamp.
HSD	honestly signifcant difference.
k_B	Boltzmann constant.
L	radiance ($W\,sr^{-1}\,m^{-2}$).
LAI	leaf area index, the ratio of projected leaf area to the ground area.
LED	light emitting diode.
LME	linear mixed effects (type of statistical model).
LSD	least significant difference.
n	number of replicates (number of experimental units per treatment).
N	total number of experimental units in an experiment.
N_A	Avogadro constant (also called Avogadro's number).
NIST	National Institute of Standards and Technology (U.S.A.).
NLME	non-linear mixed effects (statistical model).
OTC	open-top chamber.
PAR	photosynthetically active radiation, 400–700 nm. measured as energy or photon irradiance.
PC	polycarbonate, a plastic.
PG	UV action spectrum for plant growth.
PHIN	UV action spectrum for photoinhibition of isolated chloroplasts.
PID	proportional-integral-derivative (control algorithm).
PMMA	polymethylmethacrylate.
PPFD	photosynthetic photon flux density, another name for PAR photon irradiance (Q_{PAR}).
PTFE	polytetrafluoroethylene.
PVC	polyvinylchloride.
q	energy in one photon ('energy of light').
q'	energy in one mole of photons.
Q	photon irradiance ($\mu mol\,m^{-2}\,s^{-1}$).
$Q(\lambda)$	spectral photon irradiance ($\mu mol\,m^{-2}\,s^{-1}\,nm^{-1}$).
r_0	distance from sun to earth.
RAF	radiation amplification factor (nondimensional).
RH	relative humidity (%).
s	energy effectiveness (relative units).
$s(\lambda)$	spectral energy effectiveness (relative units).
s^p	quantum effectiveness (relative units).
$s^p(\lambda)$	spectral quantum effectiveness (relative units).
s.d.	standard deviation.
SDK	software development kit.
s.e.	standard error of the mean.
SR	spectroradiometer.
t	time.
T	temperature.
TUV	tropospheric UV.
U	electric potential difference or voltage (e.g. sensor output in V).
UV	ultraviolet radiation ($\lambda = 100$–400 nm).
UV-A	ultraviolet-A radiation ($\lambda = 315$–400 nm).
UV-B	ultraviolet-B radiation ($\lambda = 280$–315 nm).
UV-C	ultraviolet-C radiation ($\lambda = 100$–280 nm).

UV^{BE}	biologically effective UV radiation.
UTC	coordinated universal time, replaces GMT in technical use.
VIS	radiation visible to the human eye ($\approx$ 400–700 nm).
WMO	World Meteorological Organization.
VPD	water vapour pressure deficit (Pa).
WOUDC	World Ozone and Ultraviolet Radiation Data Centre.

Preface

In this handbook we discuss methods relevant to research on the responses of plants to ultraviolet (UV) radiation. We also summarize the knowledge needed to make informed decisions about manipulation and quantification of UV radiation, and the design of UV experiments. We give guidelines and practical recommendations for obtaining reliable and relevant data and interpretations. We cover research both on terrestrial and aquatic plants (seaweeds, marine angiosperms and freshwater higher plants are included, but microalgae are excluded from the scope of this work). We consider experimentation on ecological, eco-physiological and physiological questions.

The handbook will be most useful to early stage researchers (ESRs). However, more experienced researchers will also find information of interest. The guidelines themselves, we hope, will ensure a high and uniform standard of quality for UV research within our COST action, and the whole UV research community. We have written this text so that it is useful both for reading from cover to cover and for reference. It will also be useful as a textbook for training workshops aimed at ESRs.

Physiological and eco-physiological experiments can attempt to respond to different objective questions: (1) will a future increase in UV radiation affect growth and morphology of plants? (2) what is the effect of current UV radiation levels on plant growth and morphology? (3) what are the mechanisms by which plants respond to UV radiation? Ecological experiments can have other objectives, e.g. (1) does UV radiation in sunlight affect plant fitness? (2) does a differential effect of UV radiation between plant species affect the outcome of competition? (3) does the exposure to UV radiation alter plant-pathogen and plant-herbivore interactions? Finally applied research related to agricultural and horticultural production and produce is based on questions like: (1) can manipulations of UV radiation be used to manage produce quality? (2) can manipulation of UV radiation replace the use of pesticides and growth regulators? The approach suitable for a given experiment will depend on its objectives.

When doing experiments with terrestrial plants, the medium surrounding the stems and leaves is air. At short path lengths air has little influence on UV irradiance and only when considering the whole depth of the atmosphere, its UV transmittance needs to be taken into account. In contrast, water and impurities like dissolved organic matter (DOM) absorb UV radiation over relatively short path lengths, which means that in water bodies UV irradiance decreases with depth. Basic concepts of photobiology, radiation physics and UV in the natural environment of plants are discussed in chapter 1.

Varied approaches are used in the study of the effects of UV radition on plants. The main dichotomy is whether (1) UV radiation is added by means of special lamps to either sunlight or to visible light from other lamps, or (2) UV radiation in sunlight is excluded or attenuated by means of filters. Both approaches are extensively discussed in chapter 2.

For any experimental approach used in UV research we need to quantify UV radiation and express it as meaningful physical quantities that allow comparison among experiments and to natural conditions. When comparing UV irradiance from sources differing in spectral composition, the comparison requires the calculation of biologically effective doses. Quantification of UV radiation is discussed in chapter 3. The appendices present in detail the calculations needed when measuring action spectra, and for calculating biologically effective UV doses both with Excel and R. An R package which facilitates such calculations accompanies this handbook, and will be made available through CRAN (the Comprehensive R Archive Network) and the handbook's web pages at `http://uv4growth.dyndns.org`.

Both for terrestrial and aquatic plants the enclosing materials should be carefully chosen based on their UV transmittance and UV reflectance properties. This is crucial in UV research, but also in any other research with plants using an enclosing structure such as open-top chambers (OTC), greenhouses or aquaria. These and many other considerations about the cultivation of plants are discussed in chapter 4.

Only experiments well designed from the statistical point of view, allow valid conclusions to be reached. In addition a valid statistical analysis of the data, consistent with the design of the experiment and based on as few assumptions as possible, is required. Well designed experiments are also efficient in the use of resources (both time and money). The design of UV experiments and the analysis of the data obtained are discussed in chapter 5.

Finally a few words about terminology. As the same

quantities and units are used for measuring visible, and ultraviolet radiation, throughout the book we use the word "radiation" to refer to both visible and ultraviolet radiation. We prefer "radiation" to "light", since light is sometimes, but not always, used for just the portion of the electromagnetic spectrum visible to humans.

In the PDF file all links and crossreferences are 'live': just click on them to navigate through the file. They are marked by coloured boxes in the viewer but these boxes are not printed. In the list of references DOIs and URLs are also hyperlinked.

If you find mistakes, or difficult to understand passages, or have suggestions on how to improve this handbook, please, send feedback directly to the lead editor at `mailto:pedro.aphalo@helsinki.fi`?

`subject=UVHandbookEdition01`.

The PDF file can be freely distributed and the latest version will be available from the handbook web page at `http://uv4growth.dyndns.org/`. Printed copies can be obtained from `http://www.amazon.co.uk`, `http://www.amazon.de` or `http://www.amazon.com`.

Helsinki,	*Pedro J. Aphalo*
München,	*Andreas Albert*
Lund,	*Lars Olof Björn*
Edinburgh,	*Andy McLeod*
Helsinki,	*T. Matthew Robson*
Copenhagen,	*Eva Rosenqvist*
	October 2012

Acknowledgements

The writing and publication of this book was made possible by COST Action FA0906 'UV4growth'. This book is a collaborative effort of all members of the technical group on UV technology of this action, plus four authors not participating in the Action. The first conference and workshop organized by the Action in Szeged, Hungary, put the authors in contact as well as allowing them to realise that a book on UV research methods was needed. Some of the authors met again in Denmark, and spent two and a half days of intense writing and discussions thanks to the hospitality of Eva Rosenqvist and Carl-Otto Ottosen. We thank Profs. Åke Strid and Donat Häder for reading the whole manuscript and giving numerous suggestions for improvement.

A preprint of this handbook was used in a training school organised by the COST action at the University of Málaga (16–18 April, 2012). Corrections of errors, suggestions for improvement and complains about difficult to understand passages from participants are acknowledged.

We thank **Avantes** (The Netherlands), **BioSense** (Germany), **Biospherical Instruments Inc.** (U.S.A.), **Delta-T Devices Ltd.** (U.K.), **EIC (Equipos Intrumentación y Control)** (Spain), **Gooch & Housego** (U.S.A.), **Kipp & Zonen B.V.** (The Netherlands), **Ocean-Optics** (The Netherlands), **Valoya Oy** (Finland) **Skye Instruments Ltd.** (U.K.), **TriOS Mess- und Datentechnik GmbH** (Germany) and **Yankee Environmental Systems, Inc.** (U.S.A.) for providing illustrations. We thank Prof. Donat Häder for supplying the original data used to draw two figures and photographs of the ELDONET instrument. We thank Dr. Ulf Riebesell and Jens Christian Nejstgaard for photographs.

This work was funded by COST. Pedro J. Aphalo acknowledges the support of the Academy of Finland (decisions 116775 and 252548). Félix López Figueroa acknowledges the support by the Ministry of Innovation and Science of Spain (Project CGL08-05407-C03-01). Andy McLeod acknowledges the support of a Royal Society Leverhulme Trust Senior Research Fellowship and research awards from the Natural Environment Research Council (U.K.). Iván Gómez and Pirjo Huovinen acknowledge the financial support by CONICYT (Chile) through grants Fondecyt 1090494, 1060503 and 1080171.

1 Introduction

Pedro J. Aphalo, Andreas Albert, Lars Olof Björn, Lasse Ylianttila, Félix López Figueroa, Pirjo Huovinen

1.1 Research on plant responses to ultraviolet radiation

Plants are exposed to ultraviolet (UV) radiation in their natural habitats. The amount and quality of UV radiation they are exposed to depends on the time of the year, the latitude, the elevation, position in the canopy, clouds and aerosols, and for aquatic plants the depth, solutes and particles contained in the water (see sections 1.4 and 1.6). Ultraviolet radiation is consequently a carrier of information about the environment of plants. However, when exposed to enhanced doses of UV radiation or UV radiation of short wavelengths, plants can be damaged. When exposed to small doses of UV-B radiation plants respond by a mechanism involving the perception of the radiation through a photoreceptor called UVR8 (Christie et al., 2012; Heijde and Ulm, 2012; Jenkins, 2009; Rizzini et al., 2011; D. Wu et al., 2012; M. Wu et al., 2011). This protein behaves as a pigment at the top of a transduction chain that regulates gene expression. Several genes have been identified as regulated by UV-B radiation perceived through UVR8. Some are related to the metabolism of phenolic compounds and are involved in the accumulation of these metabolites.[1] However, these are not the only genes regulated by UVR8. Genes related to hormone metabolism are also affected, and this could be one of the mechanisms for photomorphogenesis by UV-B radiation, for example an increase in leaf thickness or reduction in height of plants. Morphological effects of UV-B mediated by UVR8 have been described (Wargent et al., 2009).

The irradiance of UV-A in sunlight is larger than the irradiance of UV-B and plants also have photoreceptors that absorb both UV-A radiation and blue light. The best studied of these photoreceptors are cryptochromes and phototropins. Cryptochromes are involved in many photomorphogenic responses, including the accumulation of pigments. Phototropins are well known for their role in plant movements such as stomatal opening in blue light and the movement of chloroplasts (see Christie, 2007; Möglich et al., 2010; Shimazaki et al., 2007, for recent reviews).

The balance between the different wavebands, UV-B, UV-A and PAR (photosynthetically active radiation, 400–700 nm), has a big influence on the effect of UV-B radiation on plants. Unrealistically low levels of UV-A radiation and PAR enhance the effects of UV-B (e.g. Caldwell et al., 1994). One reason for this is that UV-A radiation is required for photoreactivation, the repair of DNA damage in the light.

From the 1970's until the 1990's the main interest in research on the effects of UV-B on plants and other organisms was generated by the increase in ambient UV-B exposure caused by ozone depletion in the stratosphere (e.g. Caldwell, 1971; Caldwell and Flint, 1994b; Caldwell et al., 1989; Tevini, 1993). This led to many studies on the effects of increased UV-B radiation, both outdoors, in greenhouses and in controlled environments. Frequently the results obtained in outdoor experiments differed from those obtained indoors. This lead to the realization that it is important to use realistic experimental conditions with respect to UV-B radiation and its ratio compared to other bands of the solar spectrum. Interactions of responses to UV-B radiation with other environmental factors like availability of mineral nutrients, water and temperature, were also uncovered. Effects on terrestrial and aquatic ecosystems of ozone depletion, and the concomitant increase in UV-B radiation, have been periodically reviewed in UNEP (2011, and earlier reports). These reports include chapters on terrestrial ecosystems (Ballaré et al., 2011) and aquatic ecosystems (Häder et al., 2011).

From the 1990's onwards, the interest in the study of

---

[1] Many of these phenolics absorb UV radiation, so when they accumulate in the epidermis, they act as an UV shield (see Julkunen-Tiitto et al., 2005; Schreiner et al., 2012, for recent reviews). Other phenolics may behave as antioxidants (Julkunen-Tiitto et al., 2005; Schreiner et al., 2012).

the effects of normal (i.e. without stratospheric ozone depletion), as opposed to enhanced UV radiation increased markedly (e.g. Aphalo, 2003; Jansen and Bornman, 2012; Paul, 2001). This was in part due to the realization that even low UV exposures elicit plant responses, and that these are important for the acclimation of plants to their normal growth environment. Furthermore, as these effects were characterized, interest developed in their possible applications in agriculture and especially horticulture (e.g. Paul et al., 2005).

A further subject of current interest is the enhanced release of greenhouse gases from green and dead biomass caused by action of UV radiation on pectins (e.g. Bloom et al., 2010; Messenger et al., 2009). Another longstanding subject of research are the direct and indirect effects of solar UV radiation on litter decomposition (e.g. Austin and Ballaré, 2010; Newsham et al., 1997, 2001).

To be able to obtain reliable results from experiments on the effects of UV radiation on plants, there are many different problems that need to be addressed. This requires background knowledge of both photobiology, radiation physics, and UV climatology.

1.2 The principles of photochemistry

Light is electromagnetic radiation of wavelengths to which the human eye, as well as the photosynthetic apparatus, is sensitive ($\lambda \approx 400$ to 700 nm). However, sometimes the word *light* is also used to refer to other nearby regions of the spectrum: ultraviolet (shorter wavelengths than visible light) and infra-red (longer wavelengths). Both particle and wave attributes of radiation are needed for a complete description of its behaviour. Light particles or quanta are called photons.

Sensing of visible and UV radiation by plants and other organisms starts as a photochemical event, and is ruled by the basic principles of photochemistry:

Grotthuss law Only radiation that is actually absorbed can produce a chemical change.

Stark-Einstein law Each absorbed quantum activates only one molecule.

As electrons in molecules can have only discrete energy levels, only photons that provide a quantity of energy adequate for an electron to 'jump' to another possible energetic state can be absorbed. The consequence of this is that substances have colours, i.e. they absorb photons with only certain energies. See Nobel (2009) and Björn (2007) for detailed descriptions of the interactions between light and matter.

1.3 Physical properties of ultraviolet and visible radiation

In a physical sense, ultraviolet (UV) and visible (VIS) radiation (i.e. also PAR) are electromagnetic waves and are described by the Maxwell's equations.[2] The wavelength ranges of UV and visible radiation and their usual names are listed in Table 1.1. The long wavelengths of solar radiation, called infrared (IR) radiation, are also listed. The colour ranges indicated in Table 1.1 are an approximation. The electromagnetic spectrum is continuous with no clear boundaries between one colour and the next. Especially in the IR region the subdivision is somewhat arbitrary and the boundaries used in the literature vary. Radiation can also be thought of as composed of quantum particles or photons. The energy of a quantum of radiation in a vacuum, q, depends on the wavelength, λ, or frequency[3], ν,

$$q = h \cdot \nu = h \cdot \frac{c}{\lambda} \qquad (1.1)$$

with the Planck constant $h = 6.626 \times 10^{-34}$ J s and speed of light in vacuum $c = 2.998 \times 10^{8}$ m s^{-1}. When dealing with numbers of photons, the equation (1.1) can be extended by using Avogadro's number $N_A = 6.022 \times 10^{23}$ mol^{-1}. Thus, the energy of one mole of photons, q', is

$$q' = h' \cdot \nu = h' \cdot \frac{c}{\lambda} \qquad (1.2)$$

with $h' = h \cdot N_A = 3.990 \times 10^{-10}$ J s mol^{-1}. Example 1: red light at 600 nm has about 200 kJ mol^{-1}, therefore, 1 μmol photons has 0.2 J. Example 2: UV-B radiation at 300 nm has about 400 kJ mol^{-1}, therefore, 1 μmol photons has 0.4 J. Equations 1.1 and 1.2 are valid for all kinds of electromagnetic waves.

When a beam or the radiation passing into a space or sphere is analysed, two important parameters are necessary: the distance to the source and the measuring position—i.e. if the receiving surface is perpendicular to the beam or not. The geometry is illustrated in Figure 1.1 with a radiation source at the origin. The radiation is received at distance r by a surface of area dA, tilted by an angle α to the unit sphere's surface element, so called solid angle, $d\Omega$, which is a two-dimensional angle in a space. The relation between dA and $d\Omega$ in spherical coordinates is geometrically explained in Figure 1.1.

The solid angle is calculated from the zenith angle θ and azimuth angle ϕ, which denote the direction of the radiation beam

$$d\Omega = d\theta \cdot \sin\theta d\phi \qquad (1.3)$$

---

[2]These equations are a system of four partial differential equations describing classical electromagnetism.

[3]Wavelength and frequency are related to each other by the speed of light, according to $\nu = c/\lambda$ where c is speed of light in vacuum. Consequently there are two equivalent formulations for equation 1.1.

Table 1.1: Regions of the electromagnetic radiation associated with colours, after Iqbal (1983) and Eichler et al. (1993) with alterations.

Colour	Wavelength (nm)	Frequency (THz)
UV-C	100 – 280	3000 – 1070
UV-B	280 – 315	1070 – 950
UV-A	315 – 400	950 – 750
violet	400 – 455	750 – 660
blue	455 – 492	660 – 610
green	492 – 577	610 – 520
yellow	577 – 597	520 – 502
orange	597 – 622	502 – 482
red	622 – 700	482 – 428
far red	700 – 770	428 – 390
near IR	770 – 3000	390 – 100
mid IR	3000 – 50000	100 – 6
far IR	50000 – 10^6	6 – 0.3

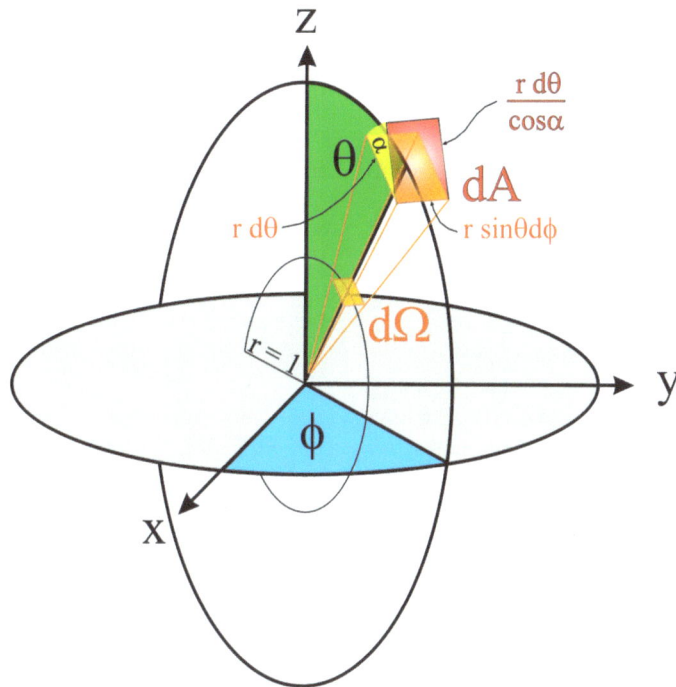

Figure 1.1: Definition of the solid angle dΩ and the geometry of areas in the space (redrawn after Eichler et al., 1993), where the given solid angle dΩ remains the same, regardless of distance r, while the exposed area exemplified by dA will change with distance r from the origin (light source) and the angle α, if the exposed area (or detector) is tilted. The angle denoted by ϕ is the azimuth angle and θ is the zenith angle.

The area of the receiving surface is calculated by a combination of the solid angle of the beam, the distance r from the radiation source and the angle α of the tilt:

$$dA = \frac{r d\theta}{\cos \alpha} \cdot r \sin \theta d\phi \qquad (1.4)$$

which can be rearranged to

$$\Rightarrow dA = \frac{r^2}{\cos \alpha} \, d\Omega \qquad (1.5)$$

Thus, the solid angle is given by

$$\Omega = \int_A \frac{dA \cdot \cos \alpha}{r^2} \qquad (1.6)$$

The unit of the solid angle is a steradian (sr). The solid angle of an entire sphere is calculated by integration of equation (1.3) over the zenith (θ) and azimuth (ϕ) angles, $0 \leq \theta \leq \pi(180°)$ and $0 \leq \phi \leq 2\pi(360°)$, and is 4π sr. For example, the sun or moon seen from the Earth's surface appear to have a diameter of about 0.5° which corresponds to a solid angle element of about 6.8×10^{-5} sr.

The processes responsible for the variation of the radiance $L(\lambda, \theta, \phi)$ as the radiation beam travels through any kind of material, are primarily absorption a and scattering b, which are called inherent optical properties, because they depend only on the characteristics of the material itself and are independent of the light field. Radiance is added to the directly transmitted beam, coming from different directions, due to elastic scattering, by which a photon changes direction but not wavelength or energy level. An example of this is Raleigh scattering in very small particles, which causes the scattering of light in a rainbow. A further gain of radiance into the direct path is due to inelastic processes like fluorescence, where a photon is absorbed by the material and reemitted as a photon with a longer wavelength and lower energy level, and Raman scattering. The elastic and inelastic scattered radiance is denoted as L^E and L^I, respectively. Internal sources of radiances, L^S, like bioluminescence of biological organisms or cells contribute also to the detected radiance. The path of the radiance through a thin horizontal layer with thickness $dz = z_1 - z_0$ is shown schematically in Figure 1.2.

Putting all this together, the radiative transfer equation is

$$\cos \theta \frac{dL}{dz} = -(a + b) \cdot L + L^E + L^I + L^S \qquad (1.7)$$

The dependencies of L on λ, θ, and ϕ are omitted here for brevity. No exact analytical solution to the radiative transfer equation exists, hence it is necessary either to use numerical models or to make approximations and find an analytical parameterisation. A numerical model

is for example the Monte Carlo method. The parameters of the light field can be simulated by modelling the paths of photons. For an infinite number of photons the light field parameters reach their exact values asymptotically. The advantage of the Monte Carlo method is a relatively simple structure of the program, and it simulates nature in a straightforward way, but its disadvantage is the time-consuming computation involved. Details of the Monte Carlo method are explained for example by Prahl et al. (1989), Wang et al. (1995)[4], or Mobley (1994).

The other way to solve the radiative transfer equation is through the development of analytical parameterisations by making approximations for all the quantities needed. In this case, the result is not exact, but it has the advantage of fast computing and the analytical equations can be inverted just as fast. This leads to the idealised case of a source-free ($L^S = 0$) and non-scattering media, i.e. $b = 0$ and therefore $L^E = L^I = 0$. Then, equation 1.7 can be integrated easily and yields

$$L(z_1) = L(z_0) \cdot e^{-\frac{a \cdot (z_1 - z_0)}{\cos \theta}} \qquad (1.8)$$

The boundary value $L(z_0)$ is presumed known. This result is known as Beer's law (or Lambert's law, Bouguer's law, Beer-Lambert law), denotes any instance of exponential attenuation of light and is exact only for purely absorbing media—i.e. media that do not scatter radiation. It is of direct application in analytical chemistry, as it describes the direct proportionality of absorbance (A) to the concentration of a coloured solute in a transparent solvent.

Different physical quantities are used to describe the "amount of radiation" and their definitions and abbreviations are listed in Table 1.2. Taking into account Equation 1.6 and assuming a homogenous flux, the important correlation between irradiance E and intensity I is

$$E = \frac{I \cdot \cos \alpha}{r^2} \qquad (1.9)$$

The irradiance decreases by the square of the distance to the source and depends on the tilt of the detecting surface area. This is valid only for point sources. For outdoor measurements the sun can be assumed to be a point source. For artificial light sources simple LEDs (light-emitting diodes) without optics on top are also effectively point sources. However, LEDs with optics—and other artificial light sources with optics or reflectors designed to give a more focused dispersal of the light—deviate to various extents from the rule of a decrease of irradiance proportional to the square of the distance from the light source.

Besides the physical quantities used for all electromagnetic radiation, there are also equivalent quantities to

---

[4]Their program is available from the website of Oregon Medical Laser Center at http://omlc.ogi.edu/software/mc/

Incoming radiance
$L(\theta, \phi)$

z_0

Loss by absorption a
and scattering b

Gain by elastic
scattered radiance
L^E into the path

Gain by inelastic scattered
radiance L^I and internal
sources L^S into the path

z_1

Outgoing radiance
$L(\theta, \phi)$

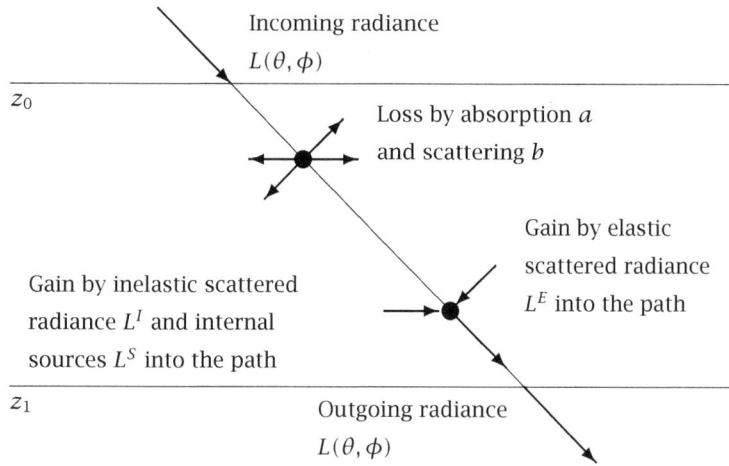

Figure 1.2: Path of the radiance and influences of absorbing and scattering particles in a thin homogeneous horizontal layer of air or water. The layer is separated from other layers of different characteristics by boundary lines at height z_0 and z_1.

Table 1.2: Physical quantities of light.

Symbol	Unit	Description
$\Phi = \frac{\partial q}{\partial t}$	$W = J\,s^{-1}$	Radiant flux: absorbed or emitted energy per time interval
$H = \frac{\partial q}{\partial A}$	$J\,m^{-2}$	Exposure: energy towards a surface area. (In plant research this is called usually *dose* (H), while in Physics *dose* refers to absorbed radiation.)
$E = \frac{\partial \Phi}{\partial A}$	$W\,m^{-2}$	Irradiance: flux or radiation towards a surface area, radiant flux density
$I = \frac{\partial \Phi}{\partial \Omega}$	$W\,sr^{-1}$	Radiant intensity: emitted radiant flux of a surface area per solid angle
$\epsilon = \frac{\partial \Phi}{\partial A}$	$W\,m^{-2}$	Emittance: emitted radiant flux per surface area
$L = \frac{\partial^2 \Phi}{\partial \Omega (\partial A \cdot \cos \alpha)} = \frac{\partial I}{\partial A \cdot \cos \alpha}$	$W\,m^{-2}\,sr^{-1}$	Radiance: emitted radiant flux per solid angle and surface area depending on the angle between radiant flux and surface perpendicular

Box 1.1: Photometric quantities

In contrast to (spectro-)radiometry, where the energy of any electromagnetic radiation is measured in terms of absolute power (J s = W), photometry measures light as perceived by the human eye. Therefore, radiation is weighted by a luminosity function or visual sensitivity function describing the wavelength dependent response of the human eye. Due to the physiology of the eye, having rods and cones as light receptors, different sensitivity functions exist for the day (photopic vision) and night (scotopic vision), $V(\lambda)$ and $V'(\lambda)$, respectively. The maximum response during the day is at $\lambda = 555$ nm and during night at $\lambda = 507$ nm. Both response functions (normalised to their maximum) are shown in the figure below as established by the Commission Internationale de l'Éclairage (CIE, International Commission on Illumination, Vienna, Austria) in 1924 for photopic vision and 1951 for scotopic vision (Schwiegerling, 2004). The data are available from the Colour and Vision Research Laboratory at `http://www.cvrl.org`. Until now, $V(\lambda)$ is the basis of all photometric measurements.

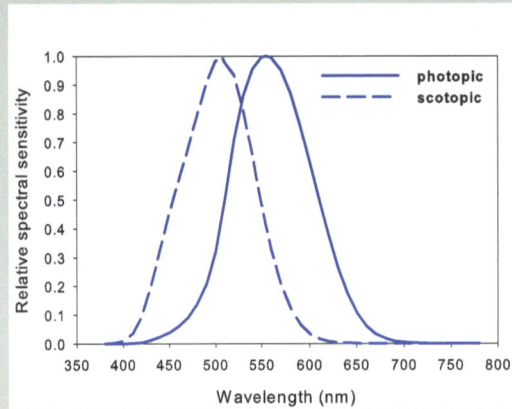

Figure. Relative spectral intensity of human colour sensation during day (solid line) and night (dashed line), $V(\lambda)$ and $V'(\lambda)$ respectively.

Corresponding to the physical quantities of radiation summarized in the table 1.2, the equivalent photometric quantities are listed in the table below and have the subscript v. The ratio between the (physiological) luminous flux Φ_v and the (physical) radiant flux Φ is the (photopic) photometric equivalent $K(\lambda) = V(\lambda) \cdot K_m$ with $K_m = 683$ lm W^{-1} (lumen per watt) at 555 nm. The dark-adapted sensitivity of the eye (scotopic vision) has its maximum at 507 nm with 1700 lm W^{-1}. The base unit of luminous intensity is candela (cd). One candela is defined as the monochromatic intensity at 555 nm (540 THz) with $I = \frac{1}{683}$ W sr^{-1}. The luminous flux of a normal candle is around 12 lm. Assuming a homogeneous emission into all directions, the luminous intensity is about $I_v = \frac{12\,\text{lm}}{4\pi\,\text{sr}} \approx 1$ cd.

Table. Photometric quantities of light.

Symbol	Unit	Description
q_v	lm s	Luminous energy or quantity of light
$\Phi_v = \frac{\partial q_v}{\partial t}$	lm	Luminous flux: absorbed or emitted luminous energy per time interval
$I_v = \frac{\partial \Phi_v}{\partial \Omega}$	cd = lm sr^{-1}	Luminous intensity: emitted luminous flux of a surface area per solid angle
$E_v = \frac{\partial \Phi_v}{\partial A}$	lux = lm m^{-2}	Illuminance: luminous flux towards a surface area
$\epsilon_v = \frac{\partial \Phi_v}{\partial A}$	lux	Luminous emittance: luminous flux per surface area
$H_v = \frac{\partial q_v}{\partial A}$	lux s	Light exposure: quantity of light towards a surface area
$L_v = \frac{\partial^2 \Phi_v}{\partial \Omega (\partial A \cdot \cos\alpha)} = \frac{\partial I_v}{\partial A \cdot \cos\alpha}$	cd m^{-2}	Luminance: luminous flux per solid angle and surface area depending on the angle between luminous flux and surface perpendicular

Box 1.2: Photon or quantum quantities of radiation.

When we are interested in photochemical reactions, the most relevant radiation quantities are those expressed in photons. The reason for this is that, as discussed in section 1.2 on page 2, molecules are excited by the absorption of certain fixed amounts of energy or quanta. The surplus energy "decays" by non-photochemical processes. When studying photosynthesis, where many photons of different wavelengths are simultaneously important, we normally use photon irradiance to describe amount of PAR. The name photosynthetic photon flux density, or PPFD, is also frequently used when referring to PAR photon irradiance. When dealing with energy balance of an object instead of photochemistry, we use (energy) irradiance. In meteorology both UV and visible radiation, are quantified using energy-based quantities. When dealing with UV photochemistry as in responses mediated by UVR8, an UV-B photoreceptor, the use of quantum quantities is preferred. According to the physical energetic quantities in the table 1.2, the equivalent photon related quantities are listed in the table below and have the subscript p.

Table. Photon quantities of light.

Symbol	Unit	Description
Φ_{p}	s^{-1}	Photon flux: number of photons per time interval
$Q = \frac{\partial \Phi_{\mathrm{p}}}{\partial A}$	$\mathrm{m}^{-2}\,\mathrm{s}^{-1}$	Photon irradiance: photon flux towards a surface area, photon flux density (sometimes also symbolised by E_{p})
$H_{\mathrm{p}} = \int_t Q \, \mathrm{d}t$	m^{-2}	Photon exposure: number of photons towards a surface area during a time interval, photon fluence

These quantities can be also used based on a 'chemical' amount of moles by dividing the quantities by Avogadro's number $N_A = 6.022 \times 10^{23} \, \mathrm{mol}^{-1}$. To determine a quantity in terms of photons, an energetic quantity has to be weighted by the number of photons, i.e. divided by the energy of a single photon at each wavelength as defined in equation 1.1. This yields for example

$$\Phi_{\mathrm{p}} = \frac{\lambda}{h\,c} \cdot \frac{\partial q}{\partial t} \qquad \text{and} \qquad Q(\lambda) = \frac{\lambda}{h\,c} \cdot E(\lambda)$$

When dealing with bands of wavelengths, for example an integrated value like PAR from 400 to 700 nm, it is necessary to repeat these calculations at each wavelength and then integrate over the wavelengths. For example, the PAR photon irradiance or PPFD in moles of photons is obtained by

$$\mathrm{PPFD} = \frac{1}{N_A} \int_{400\,\mathrm{nm}}^{700\,\mathrm{nm}} \frac{\lambda}{hc} \, E(\lambda) \, \mathrm{d}\lambda$$

For integrated values of UV-B or UV-A radiation the calculation is done analogously by integrating from 280 to 315 nm or 315 to 400 nm, respectively.

If we have measured (energy) irradiance, and want to convert this value to photon irradiance, the exact conversion will be possible only if we have information about the spectral composition of the measured radiation. Conversion factors at different wavelengths are given in the table below. For PAR, $1 \, \mathrm{W\,m}^{-2}$ of "average daylight" is approximately $4.6 \, \mu\mathrm{mol\,m}^{-2}\,\mathrm{s}^{-1}$. This is exact only if the radiation is equal from 400 to 700 nm, because the factor is the value at the central wavelength at 550 nm. Further details are discussed in section 3.1 on page 71.

Table. Conversion factors of photon and energy quantities at different wavelengths.

	$\mathrm{W\,m}^{-2}$ to $\mu\mathrm{mol\,m}^{-2}\,\mathrm{s}^{-1}$	λ (nm)
	2.34	280
UV-B	2.49	298
	2.63	315
UV-A	2.99	358
	3.34	400
PAR	4.60	550
	5.85	700

describe visible radiation, so called photometric quantities. The human eye as a detector led to these photometric units, and they are commonly used by lamp manufacturers to describe their artificial light sources. See Box 1.1 on page 6 for a short description of these quantities and units.

There are, in principle, two possible approaches to measuring radiation. The first is to observe light from one specific direction or viewing angle, which is the radiance L. The second is to use a detector, which senses radiation from more than one direction and measures the so-called irradiance E of the entire sphere or hemisphere. The correlation between irradiance E and radiance L of the wavelength λ is given by integrating over all directions of incoming photons.

$$E_0(\lambda) = \int_\Omega L(\lambda, \Omega) d\Omega \qquad (1.10)$$

$$E(\lambda) = \int_\Omega L(\lambda, \Omega) |\cos \alpha| d\Omega \qquad (1.11)$$

Depending on the shape of a detector (which may be either planar or spherical) the irradiance is called (plane) irradiance E or fluence rate (also called scalar irradiance) E_0. A planar sensor detects incoming photons depending on the incident angle and a spherical sensor detects all photons equally weighted for all directions. See section 3.1 on page 71 for a more detailed discussion.

Here we have discussed the properties of light based on energy quantities. In photobiology there are good reasons to quantify radiation based on photons. See Box 1.2 on page 7, and section 3.1 on page 71.

1.4 UV in solar radiation

When dealing with solar radiation, we frequently need to describe the position of the sun. The azimuth angle (ϕ) is measured clockwise from the North on a horizontal plane. The position on the vertical plane is measured either as the zenith angle (θ) downwards from the zenith, or as an elevation angle (h) upwards from the horizon. Consequently $h + \theta = 90° = \frac{\pi}{2}$ radians. See Figure 1.3 for a diagram. In contrast to Figure 1.1 and the discussion in section 1.3 where the point radiation source is located at the origin of the system of coordinates, when describing the position of the sun as in Figure 1.3 the observer is situated at the origin.

Ultraviolet and visible radiation are part of solar radiation, which reaches the Earth's surface in about eight minutes (t = time, r_0 = distance sun to earth, c = velocity of light in vacuum):

$$t = \frac{r_0}{c} \approx \frac{150 \times 10^9 \text{ m}}{3 \times 10^8 \frac{\text{m}}{\text{s}}} = 500 \text{ s} = 8.3 \text{ min}$$

The basis of all passive measurements is the incoming solar radiation, which can be estimated from the known activity of the sun ('productivity of photons'), that can be approximated by the emitted spectral radiance (L_s) described by Planck's law of black body radiation at temperature T, measured in degrees Kelvin (K):

$$L_s(\lambda, T) = \frac{2hc^2}{\lambda^5} \cdot \frac{1}{e^{(hc/k_B T\lambda)} - 1} \qquad (1.12)$$

with Boltzmann's constant $k_B = 1.381 \times 10^{-23}$ JK^{-1}. The brightness temperature of the sun can be determined by Wien's displacement law, which gives the peak wavelength of the radiation emitted by a blackbody as a function of its absolute temperature

$$\lambda_{max} \cdot T = 2.898 \times 10^6 \text{ nm K} \qquad (1.13)$$

This means that for a maximum emission of the sun at about 500 nm the temperature of the sun surface is about 5800 K. The spectral irradiance of the sun $E_s(\lambda)$ can be estimated assuming a homogeneous flux and using the correlation of intensity I and radiance L from their definitions in table 1.2. The intensity of the sun $I_s(\lambda)$ is given by the radiance $L_s(\lambda)$ multiplied by the apparent sun surface (a non-tilted disk of radius $r_s = 7 \times 10^5$ km). To calculate the decreased solar irradiance at the moment of reaching the Earth's atmosphere, the distance of the sun to the Earth ($r_0 = 150 \times 10^6$ km) has to be taken into account due to the inverse square law of irradiance of equation (1.9). Thus, the extraterrestrial solar irradiance is

$$E_s(\lambda) = L_s(\lambda) \cdot \frac{\pi r_s^2}{r_0^2} \qquad (1.14)$$

Remembering the solid angle of equation (1.6), the right multiplication factor represents the solid angle of the sun's disk as seen from the Earth's surface ($\approx 6.8 \times 10^{-5}$ sr). Figure 1.4 shows the spectrum of the measured extraterrestrial solar radiation (Wehrli, 1985)[5] and the spectrum calculated by equation 1.14 using Planck's law of equation 1.12 at a black body temperature of 5800 K. Integrated over all wavelengths, E_s is about 1361 to 1362 W m^{-2} at top of the atmosphere (Kopp and Lean, 2011). This value is called the 'solar constant'. In former times, depending on different measurements, E_s varies by a few percent (Iqbal, 1983). For example, the irradiance at the top of the atmosphere (the integrated value) changes by ± 50 W m^{-2} (3.7 %) during the year due to distance variation caused by orbit excentricity (Mobley, 1994). More accurate measurements during the last 25 years by space-borne radiometers show a variability of the solar radiation of a few tenth of a percent. A detailed analysis is given by Fröhlich and Lean (2004). E_s can also be calculated by the Stefan-Boltzmann Law: the total energy emitted from the surface of a black body is proportional

---

[5] Available as ASCII file at PMODWRC, ftp://ftp.pmodwrc.ch/pub/publications/pmod615.asc

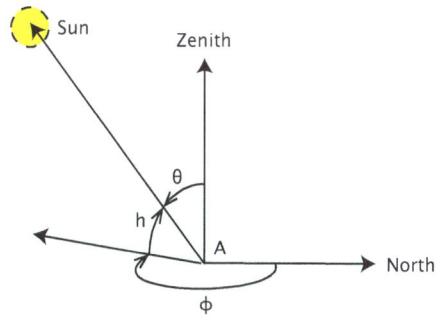

Figure 1.3: Position of the sun in the sky and the different angles used for its description by an observer located at point A. The azimuth angle is ϕ, the elevation angle is h and the zenith angle is θ. These angles are measured on two perpendicular planes, one horizontal and one vertical.

to the fourth power of its temperature. For an isotropically emitting source (Lambertian emitter), this means

$$L = \frac{\sigma}{\pi} \cdot T^4 \qquad (1.15)$$

with the Stefan-Boltzmann constant $\sigma = 5.6705 \times 10^{-8}$ $W \cdot m^{-2} \, K^{-4}$. With $T = 5800$ K equation 1.15 gives the radiance of the solar disc. From this value, we can obtain an approximation of the solar constant, by taking into account the distance from the Earth to the Sun and the apparent size of the solar disc (see equations 1.6 and 1.9).

The total solar irradiance covers a wide range of wavelengths. Using some of the 'colours' introduced in table 1.1, table 1.3 lists the irradiance and fraction of E_s of different wavelength intervals.

The extraterrestrial solar spectrum differs from that at ground level due to the absorption of radiation by the atmosphere, because the absorption peaks of water, CO_2 and other components of the atmosphere, cause corresponding valleys to appear in the solar spectrum at ground level. For example, estimates from measurements of the total global irradiance at Helmholtz Zentrum München (11.60° E, 48.22° N, 490 m above sea level) on two sunny days (17th April 1996, sun zenith angle of 38° and 27th May 2005, 27°) result in about 5% for wavelengths below 400 nm, about 45% from 400 to 700 nm, and about 50% above 700 nm. In relation to plant research, only the coarse structure of peaks and valleys is relevant, because absorption spectra of pigments *in vivo* have broad peaks and valleys. However, the solar spectrum has a much finer structure, due to emission and absorption lines of elements, which is not observable with the spectroradiometers normally used in plant research.

At the Earth's surface, the incident radiation or *global radiation* has two components, direct radiation and scattered or 'diffuse' radiation. Direct radiation is ra-

diation travelling directly from the sun, while diffuse radiation is that scattered by the atmosphere. Diffuse radiation is what gives the blue colour to the sky and white colour to clouds. The relative contribution of direct and diffuse radiation to global radiation varies with wavelength and weather conditions. The contribution of diffuse radiation is larger in the UV region, and in the presence of clouds (Figures 1.5 and 1.6).

Not only total irradiance, but also the wavelength distribution of the solar spectrum changes with the seasons of the year and time of day. The spectral wavelength distribution is also changed by the amount of UV-absorbing ozone in the atmosphere, known as the ozone column. Figure 1.7 shows how spectral irradiance changes throughout one day. When the whole spectrum is plotted using a linear scale the effect of ozone depletion is not visible, however, if we plot only the UV region (Figure 1.8) or use a logarithmic scale (Figure 1.9), the effect becomes clearly visible. In addition, on a log scale, it is clear that the relative effect of ozone depletion on the spectral irradiance at a given wavelength increases with decreasing wavelength.

Seasonal variation in UV-B irradiance has a larger relative amplitude than variation in PAR (Figure 1.10). This causes a seasonal variation in the UV-B: PAR ratio (Figure 1.11). In addition to the regular seasonal variation, there is random variation as a result of changes in clouds (Figure 1.11). Normal seasonal and spatial variation in UV can be sensed by plants, and could play a role in their adaptation to seasons and/or their position in the canopy.

UV-B irradiance increases with elevation in mountains and with decreasing latitude (Figure 1.12) and is particularly high on high mountains in equatorial regions. This has been hypothesized to be a factor in the determination of the tree line[6] in these mountains (Flenley, 1992).

---

[6] *Tree line* is the highest elevation on a mountain slope at which tree species are naturally able to grow.

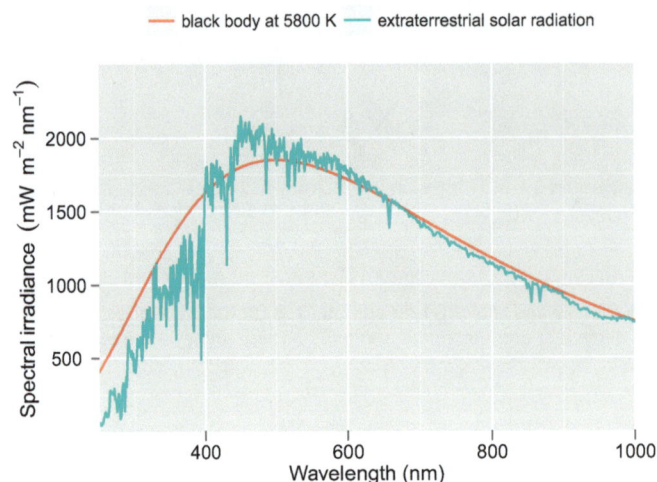

Figure 1.4: Extraterrestrial solar spectrum after Wehrli (1985) (green line) and spectrum of a black body at 5800 K (red line), calculated using Planck's law (equation 1.12) and converted to extraterrestrial spectral irradiance with equation 1.14.

Table 1.3: Distribution of the extraterrestrial solar irradiance E_s constant in different wavelength intervals calculated using the data of Wehrli (1985) shown in Figure 1.4.

Colour	Wavelength (nm)	Irradiance ($\mathrm{W\,m^{-2}}$)	Fraction of E_s (%)
UV-C	100 – 280	7	0.5
UV-B	280 – 315	17	1.2
UV-A	315 – 400	84	6.1
VIS	400 – 700	531	38.9
near IR	700 – 1 000	309	22.6
mid and far IR	> 1 000	419	30.7
total		1 367	100.0

An increase in the UV-B irradiance is caused by depletion of the ozone layer in the stratosphere, mainly as a consequence of the release of chlorofluorocarbons (CFCs), used in cooling devices such as refrigerators and air conditioners, and in some spray cans (see Graedel and Crutzen, 1993). The most dramatic manifestation of this has been the seasonal formation of an "ozone hole" over Antarctica. It is controversial whether a true ozone hole has already formed in the Arctic, but strong depletion has occurred in year 2011 (Manney et al., 2011) and atmospheric conditions needed for the formation of a "deep" ozone hole are not very different from those prevalent in recent years. Not so dramatic, but consistent, depletion has also been observed at mid-latitudes in both hemispheres. CFCs and some other halocarbons have been phased out following the Montreal agreement and later updates. However, as CFCs have a long half life in the atmosphere, of the order of 100 years, their effect on the ozone layer will persist for many years, even after their use has been drastically reduced. Model-based predictions of changes in atmospheric circulation due to global climate change have been used to derive future trends in UV index and ozone column thickness (Hegglin and Shepherd, 2009). In addition, increased cloudiness and pollution, could lead to decreased UV and PAR, sometimes called 'global dimming' (e.g. Stanhill and Cohen, 2001). It should be noted that, through reflection, broken clouds can locally increase UV irradiance to values above those under clear-sky conditions (S. B. Díaz et al., 1996; Frederick et al., 1993).

Figure 1.5: Sky photos in different portions of the light spectrum. They show that in the UV-A band the diffuse component is proportionally larger than it is at longer wavelengths. This can be seen as reduced contrast. Photographs taken by L. Ylianttila at the fortress of Suomenlinna (`http://www.suomenlinna.fi/en`), Helsinki, Finland.

1.5 UV radiation within plant canopies

The attenuation of visible and UV radiation by canopies is difficult to describe mathematically because it is a complex phenomenon. The spatial distribution of leaves is in most cases not uniform, the display angle of the leaves is not random, and may change with depth in the canopy, and even in some cases with time-of-day. Here we give only a description of the simplest approach, the use of an approximation based on Beer's law as modified by Monsi and Saeki (1953), reviewed by Hirose (2005). Beer's law (Equation 1.8) assumes a homogeneous light absorbing medium such as a solution. However, a canopy is heterogenous, with discrete light absorbing objects (the leaves and stems) distributed in a transparent medium (air).

$$I_z = I_0 \cdot e^{-KL_z} \qquad (1.16)$$

Equation 1.16 describes the radiation attenuated as a function of leaf area index (L or LAI) at a given canopy depth (z). The equation does not explicitly account for the effects of the statistical spatial distribution of leaves and the effects of changing incidence angle of the radiation. Consequently, the empirical extinction coefficient (K) obtained may vary depending on these factors. K is not only a function of plant species (through leaf optical properties, and how leaves are displayed), but also of time-of-day, and season-of-year—as a consequence of solar zenith angle—and degree of scattering of the incident radiation. As the degree of scattering depends on clouds, and also on wavelength, the extinction coefficient is different for UV and visible radiation. Radiation extinction in canopies has yet to be studied in detail with respect to UV radiation, mainly because of difficulties in the measurement of UV radiation compared to PAR, a spectral region which has been extensively studied.

Ultraviolet radiation is strongly absorbed by plant surfaces, although cuticular waxes and pubescence on leaves can sometimes increase UV reflectance. The diffuse component of UV radiation is larger than that of visible light (Figure 1.10). In sunlit patches in forest gaps the diffuse radiation percentage is lower than in open areas, because direct radiation is not attenuated but part of the sky is occluded by the surrounding forest. Attenuation with canopy depth is on average usually more gradual for UV than for PAR. The UV irradiance decreases with depth in tree canopies, but the UV:PAR ratio tends to increase (see Brown et al., 1994). In contrast, Deckmyn et al. (2001) observed a decrease in UV:PAR ratio in white clover canopies with planophyle leaves. Allen et al. (1975) modelled the UV-B penetration in plant canopies, under normal and depleted ozone conditions. Parisi and Wong (1996) measured UV-B doses within model plant canopies using dosimeters. The position of leaves affects UV-B exposure, and it has been observed that heliotropism can moderate exposure and could be a factor contributing to differences in tolerance among crop cultivars (Grant, 1998, 1999a,b, 2004).

Detailed accounts of different models describing the

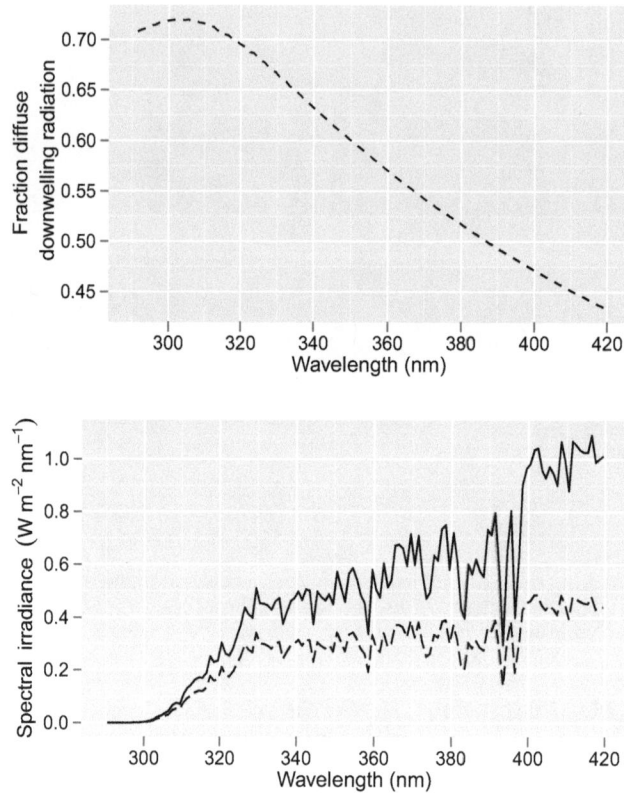

Figure 1.6: Diffuse component in solar UV. Spectral irradiance of total downwelling radiation (lower panel, solid line), diffuse downwelling radiation (lower panel, long dashes), and ratio of diffuse downwelling to total downwelling spectral irradiance (upper panel, dashed line) are shown. Data from TUV model (version 4.1) for solar zenith angle = 40°00′, cloud-free conditions, 300 Dobson units. Simulations done with the Quick TUV calculator at `http://cprm.acd.ucar.edu/Models/TUV/Interactive_TUV/`.

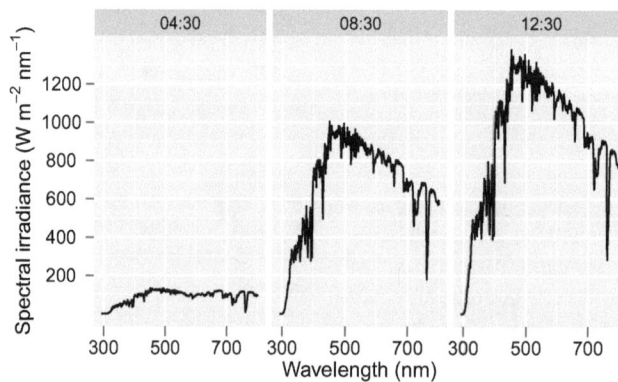

Figure 1.7: The solar spectrum through half a day. Simulations of global radiation (direct plus diffuse radiation) spectral irradiance on a horizontal surface at ground level) for a hypothetical 21 May with cloudless sky at Jokioinen (60°49′N, 23°30′E), under normal ozone column conditions. Effect of depletion is so small on the solar spectrum as a whole, that it would not visible in this figure. See Kotilainen et al. (2011) for details about the simulations.

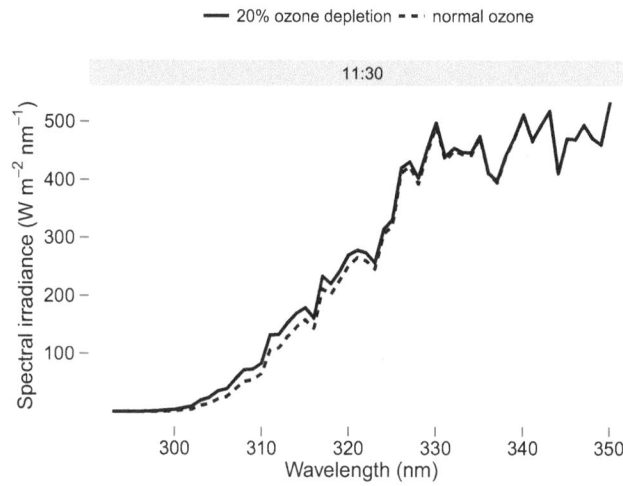

Figure 1.8: The effect of ozone depletion on the UV spectrum of global (direct plus diffuse) solar radiation at noon. See fig. 1.7 for details.

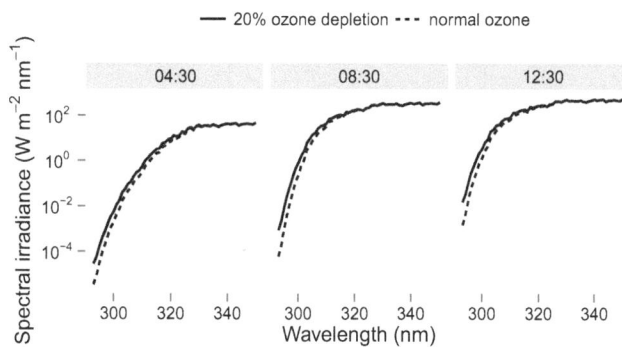

Figure 1.9: The solar UV spectrum through half a day. The effect of ozone depletion on global (direct plus diffuse) radiation. A logarithmic scale is used for spectral irradiance. See fig. 1.7 for details.

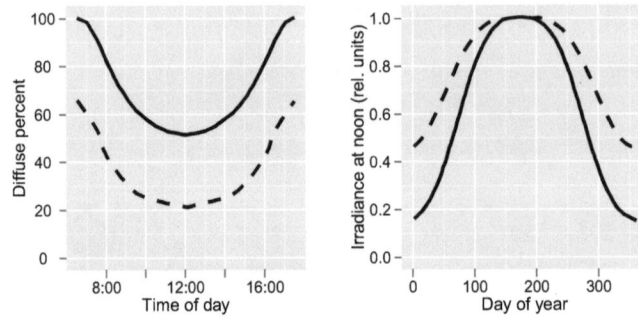

Figure 1.10: UV-B and PAR. Left: Diffuse radiation as percentage of total (direct + diffuse) radiation in the UV-B (solid line) and PAR (dashed line) wavebands for open areas in a humid temperate climate under a clear sky. In cloudy conditions the percentage of diffuse radiation increases. Day of year not specified. Redrawn from Flint and Caldwell (1998). Right: Seasonal variation in modelled, clear sky, solar-noon, UV-B (solid line) and PAR (dashed line) irradiance above the canopy for Maryland, USA. Irradiance expressed relative to annual maximum of each waveband. Adapted from Brown et al. (1994).

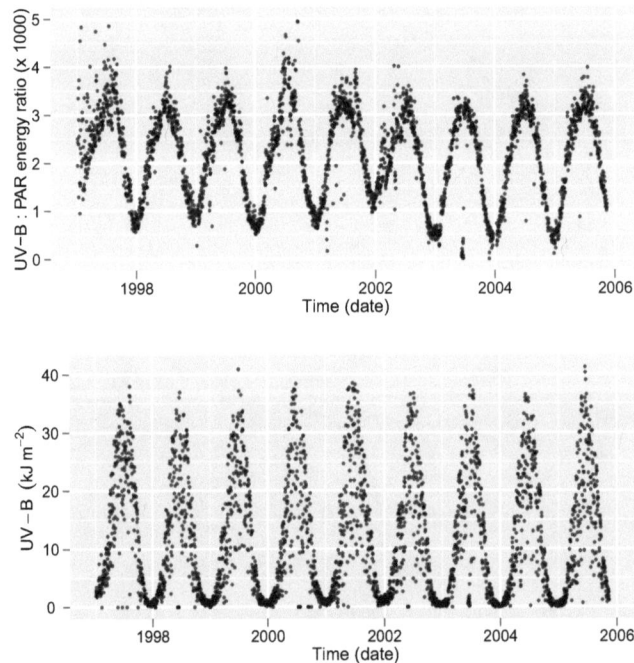

Figure 1.11: Seasonal variation in UV-B radiation at Erlangen, Germany (54° 10' N, 07° 51' E, 280 m asl). (Top) UV-B:PAR energy ratio, calculated from daily exposures, and (bottom) UV-B daily exposure, measured with ELDONET instruments (see Figure 2 in Häder et al., 2007, for details).

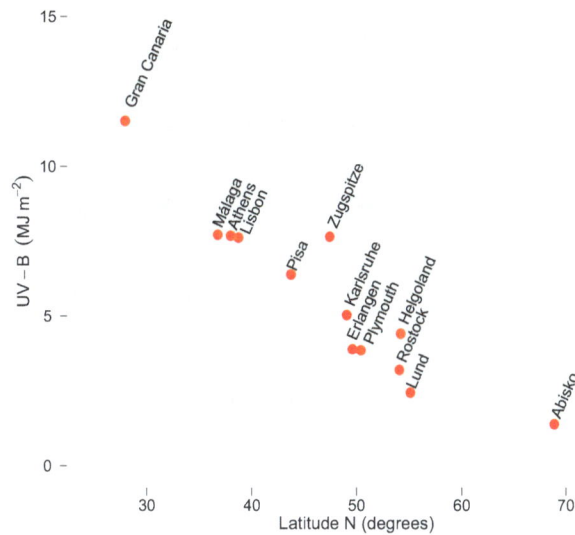

Figure 1.12: Latitudinal variation in UV-B radiation in the Northern hemisphere. UV-B annual exposure, measured with ELDONET instruments (see Häder et al., 2007, for details).

interaction of radiation and plant canopies, taking into account the properties of foliage, are given by Campbell and Norman (1998) and Monteith and Unsworth (2008).

1.6 UV radiation in aquatic environments

As solar radiation passes through a body of water, its spectrum changes with depth in a wavelength-dependent manner, determined by the optical characteristics of that water body. The penetration of UV radiation through water bodies can vary from only few centimetres in highly humic lakes (Huovinen et al., 2003; Kirk, 1994a,b) to dozens of metres in the oceans (Kirk, 1994a,b; Smith et al., 1992). Some irradiance is reflected at the water surface, but the extent to which wavelengths in the UV to IR range penetrate water bodies depends mainly on (1) attenuation by water itself, (2) coloured dissolved organic matter (CDOM), and seston. Seston is the sum of living organic material (mainly phytoplankton) and non-living material (tripton). Non-living particles are further distinguished between organic material (detritus) and inorganic suspended matter. Each fraction has its own characteristic spectral absorption and scattering properties (reviewed by Dekker, 1993; Hargreaves, 2003; Kirk, 1994a,b; Wozniak and Dera, 2007).

Particularly in coastal areas and shallow areas of lakes and streams, irradiance reflected from the ground or seabed beneath the water (henceforth bottom) influences the profile of radiation through the aquatic environment. This reflectance is described by a bidirectional reflectance distribution function (BRDF), which is wavelength specific and depends on the incident and reflected angle. If the reflectance is equally distributed in all directions, the bottom is a so called Lambertian reflector and the BRDF is constant. The bottom reflectance is greatly influenced by its slope and properties, i.e. whether the bottom is bare sediment or covered by algae and submersed vegetation (e.g. Albert and Mobley, 2003; Maritorena et al., 1994; Mobley and Sundman, 2003; Mobley et al., 2003; Pinnel, 2007).

1.6.1 Refraction

The refraction of incoming (downwelling) radiance at the water surface can be determined by Snell's law, which describes the angular refraction of the incident beam. The radiation passes the first medium with a refractive index n_1 and then the second medium with a refractive index n_2. If the incoming direction of the radiation is given by the angle θ_1, the beam is refracted to the angle θ_2. Snell's law is

$$n_1 \cdot \sin\theta_1 = n_2 \cdot \sin\theta_2 \qquad (1.17)$$

For the case of radiation arriving from the air under the incident angle θ_i and going into the water with the transmitted angle θ_t, this yields a refractive index for the air of $n_a = 1$ and for the water of $n_W = 1.33$

$$\theta_t = \arcsin(0.75 \cdot \sin\theta_i) \qquad (1.18)$$

Theoretically, n_W is not constant but depends on temperature, wavelength and salinity, as described by Quan

15

and Fry (1995). In principle, the shorter the wavelength, the higher the refractive index of water, but in practice the wavelength-dependent difference in refraction is unimportant. For example comparing the values at 400 and 800 nm for 20°C and no salinity produces a difference of < 0.5%. If the wind speed is high, the slope of surface waves also has to be taken into account. A rough surface reflects and transmits the incoming radiation beam in more directions and makes the radiation field more diffuse than a smooth surface.

1.6.2 Absorption and scattering by pure water

Water itself absorbs and scatters radiation. The optical properties of the water in the visible and ultraviolet (UV) spectrum are not precisely known, since no theoretical model exists which exactly describes the absorption and scattering properties of pure water. Therefore, it is necessary to rely on laboratory measurements to approximate the values of these parameters. Investigations into absorption by water a_W were initially documented by Morel (1974), Smith and Tyler (1976), Smith and Baker (1981), Pegau and Zaneveld (1993) and more recently by Buiteveld et al. (1994) and Hakvoort (1994). The absorption properties of water also depend on temperature. The influence of temperature is weak below 700 nm, but its effect increases with increasing wavelength; so, for example, a temperature increase of 10 K produces a $\approx 7\%$ change in the absolute value of a_W at 740 nm.

The scattering of radiation by molecules in liquids has been modelled theoretically by Smoluchowski (1908) and Einstein (1910). This approach is based on statistical thermodynamics and is called the theory of fluctuation. Theoretically, the scattering function is wavelength-dependent and follows the λ^{-4} law. Experiments show a slight deviation from the model, giving a better correlation with $\lambda^{-4.32}$ (Morel, 1974) due to the effects of isothermal compressibility, the refractive index of water, and the pressure derivative of the refractive index of water (Hakvoort, 1994).

The wavelength dependency of the absorption and scattering coefficients of pure water are shown in Figure 1.13 using data from Hakvoort (1994). Water mainly contributes to the attenuation of PAR and IR wavelengths, since absorption by pure water increases from around 550 nm towards longer wavelengths.

1.6.3 Absorption and scattering by water constituents

Absorption and scattering by water constituents is the sum of (1) absorption by CDOM, sometimes also called yellow or humic substances, gilvin or gelbstoff, (2) absorption and scattering by living material like phytoplankton,

and (3) absorption and scattering by dead organic and inorganic particles. The influence of each constituent on the scattering process depends on wavelength, particle size, concentration, and refractive index. Theoretical details are explained in, for example, Hulst (1981).

CDOM mainly refers to coloured dissolved humic materials and consists of humic and fulvic acids, originating from decomposed plant material suspended in the water or entering from the surrounding catchment area. The pigments in humic and fulvic acids absorb strongly in the blue and UV wavelengths and are dissolved and therefore do not scatter irradiance. Kalle (1966) recognised that CDOM absorption decreases exponentially with increasing wavelength in the visible part of the spectra. Following the study of Morel and Prieur (1976), Bricaud et al. (1981) expressed this relationship in the following model: for a known absorption at a wavelength $\lambda_0 = 440$ nm, the CDOM absorption a_Y can be determined by

$$a_Y(\lambda) = a_Y(\lambda_0) \cdot e^{-s_Y(\lambda - \lambda_0)} \qquad (1.19)$$

Although the exponential coefficient s_Y is variable, a standard value of $s_Y = 0.014$ nm^{-1} is commonly used. Bricaud et al. (1981) compared the value of s_Y across many data sets and reported a standard deviation of only $\Delta s_Y = \pm 0.003$ nm^{-1}. The amount of CDOM in water is determined by filtration using membrane filters of 0.2 μm pore size. The filtrate is collected into a quartz cuvette with length l and put into a (double beam) spectrophotometer to measure its absorbance (optical density) $A(\lambda)$. Then, $a_Y(\lambda) = 2.303 \cdot \frac{A(\lambda)}{l}$ (Kirk, 1994a). The absorption coefficient at 440 nm has been used as an indication of optical colour (Kirk, 1994a), while size of humic molecules has been estimated from the ratio $\frac{a_Y(\lambda=250\text{nm})}{a_Y(\lambda=365\text{nm})}$, with increasing size indicated by smaller ratios (Haan, 1972, 1993; Haan et al., 1987). To determine a_Y from clearer (e.g. oceanic) waters, a cuvette with a 10 cm pathlength is generally needed due to their low values of absorption.

Phytoplankton can contribute to the attenuation of PAR through absorption by their photosynthetic pigments such as chlorophyll and pheophytin, but they can also cause scattering. The absorption by phytoplankton a_P is the sum of absorption by each pigment multiplied each by their concentrations. Due to the fact that many species of phytoplankton occur in aquatic environments and every species contains more than one pigment, it is more practicable to calculate the absorption by mean specific absorption coefficients for each different algal species separately. This has been done by Gege (1998) for freshwater Lake Constance in Germany and by Prieur and Sathyendranath (1981) for an oceanic environment. Besides these examples, there are other models for oceanic waters that use the specific *in vivo*

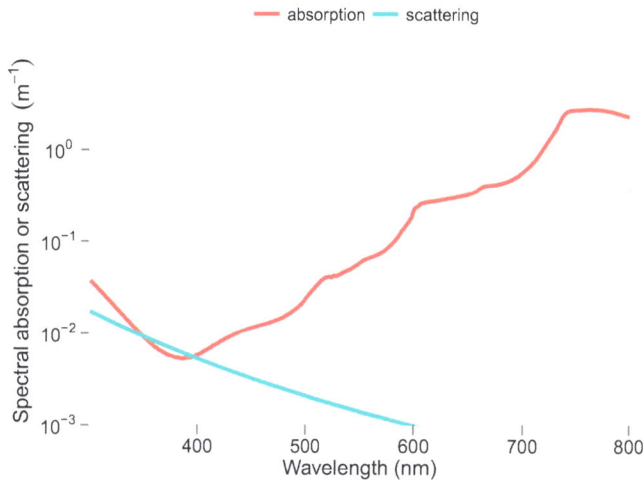

Figure 1.13: Absorption (red line) and scattering (green line) coefficients of pure water as a function of wavelength, after Hakvoort (1994). Shown in logarithmic scale.

absorption coefficient and concentration of chlorophyll-a, a^*_{chl} and C_{chl}, respectively. Morel (1991) found that the power law $a_P = 0.06\, a^*_{chl}\, [C_{chl}]^{0.65}$, which was first proposed by Prieur and Sathyendranath (1981), provided the best estimate of the absorption coefficient for his data set. C_{chl} is the concentration of chlorophyll-a in units of $\mu g\, l^{-1}$. Figure 1.14 shows the specific oceanic chlorophyll absorption coefficient from Morel (1991) normalized to maximum absorption at 440 nm. Figure 1.14 also shows laboratory measurements of chlorophyll-a and chlorophyll-b absorption[7] after Frigaard et al. (1996). Spectra for chlorophyll absorption, and that of various other photochemically-relevant substances, are also available in computer software such as PhotochemCAD[8] (Dixon et al., 2005; Du et al., 1998). The empirical model of Bricaud et al. (1995) parameterises the specific absorption coefficient from C_{chl}. This model draws on extensive studies of more than 800 spectra to give $a^*_P(\lambda) = A(\lambda)[C_{chl}]^{-B(\lambda)}$ with positive empirical coefficients A and B depending on wavelength. The model incorporates both the package effect of phytoplankton cells and the effect of the varying pigment composition on absorption.

The particulate structure of phytoplankton cells causes scattering. The influence of phytoplankton on the total scattering coefficient depends on the other constituents of the water body. For water containing a low concentration of inorganic suspended sediment scattering is driven by the concentration of phytoplankton (as occurs in the open ocean). Gordon and Morel (1983) developed an empirical model, which directly correlates scattering with the pigment concentration of chlorophyll-a C_{chl} in units of $\mu g\, l^{-1}$. The scattering coefficient of phytoplankton b_P in units of m^{-1} is given by

$$b_P(\lambda) = B \cdot [C_{chl}]^{0.62} \cdot \left(\frac{\lambda_0}{\lambda}\right) \qquad (1.20)$$

with $\lambda_0 = 550$ nm and $B = 0.3$ as mean values for oceanic waters dominated by phytoplankton. The equation (1.20) is valid for a range of phytoplankton concentrations from 0.05 to 1 $\mu g\, l^{-1}$. Gordon and Morel (1983) found that oceanic waters have a value of $B \le 0.45$. Higher values are used for other aquatic environments, for example turbid coastal waters. In coastal regions, Sathyendranath et al. (1989) proposed that the scattering coefficient is indirectly proportional to absorption by phytoplankton: $b_P(\lambda) \propto 1/a_P(\lambda)$. The proportionality factor depends on the concentration of chlorophyll-a in the same manner as the scattering coefficient of equation (1.20). Dekker (1993) investigated the contribution to scattering of each water constituent in inland waters[9]. He found that the composition of scattering particles was more variable in inland waters than in the ocean and it depended on the

---

[7]Data available from Frigaard's website at the University of Copenhagen at `http://www.bio.ku.dk/nuf/resources/scitab/chlabs/index.htm`.

[8]Latest software version available at the PhotochemCAD website at `http://photochemcad.com`. Data available at the PhotochemCAD data site at `http://omlc.ogi.edu/spectra/PhotochemCAD/index.html`.

[9]Inland waters include rivers, streams and lakes.

trophic state of the water and therefore on the distribution of organic and inorganic particulate matter. The scattering and backscattering coefficient of phytoplankton can be determined by $b_P(\lambda) = b_P^*(\lambda) \cdot C_P$ where the specific scattering coefficient of phytoplankton is b_P^*. For lakes Dekker (1993) reported that the specific scattering coefficient of phytoplankton ranges from 0.12 to 0.18 m^2mg^{-1} at a wavelength of 550 nm. The specific scattering coefficients can be obtained by integrating the scattering phase function of the observed matter, here, phytoplankton. Extensive and commonly used measurements were done by Petzold (1977). Other functions can be found for example in Mobley (1994).

Particulate matter in water bodies consists of organic and inorganic material. The organic constituents are contained in phytoplankton cells or are fragments of dead plankton and faecal pellets of zooplankton. These parts are often called detritus. Inorganic particles include suspended mineral coming from inflows or resuspension at coastal regions. They mainly consist of quartz, clay, and calcite. There are only a few published values of the specific absorption coefficients of suspended particles in water from aquatic environments because they are difficult to separate into their individual constituent parts. A comparison of these values is given by Pozdnyakov and Grassl (2003). In general, absorption by all suspended particles in most water bodies is very low and it is negligible for inorganic particles. Roesler et al. (1989) produced the following relationship for absorption by detritus in coastal waters which is very similar that of CDOM:

$$a_X(\lambda) = a_X(\lambda_0) \cdot e^{-s_X(\lambda - \lambda_0)} \qquad (1.21)$$

with a mean value of $s_X = 0.011$ nm^{-1} and $a_X(\lambda_0) = 0.09$ m^{-1} at $\lambda_0 = 400$ nm for their data. Particulate matter in general causes more scattering of irradiance than it absorbs. In coastal waters and freshwater, scattering is higher than in oceanic waters due to the additional presence of particles not related to phytoplankton. These particles come from suspended inorganic sediments of different sizes. Scattering is caused by differences in the refractive indices of the two materials (the water medium and the material of the particles) and are due to the ratio of particle size to wavelength. Different functions of scattering coefficients and phase functions are, for example, described by Mobley (1994). Especially turbid and coastal waters, or rivers and lakes, are dominated by large particles (> 1 μm and a refraction index of 1.03). Therefore, the scattering coefficient of non-living particles b_X can be estimated while neglecting their size distribution and wavelength dependence. Thus, b_X is derived from $b_X(\lambda) = b_X^* \cdot C_X$, with the concentration of the total suspended matter C_X and the specific scattering coefficients b_X^*. Dekker (1993) gave example specific scattering coef-

ficients of 0.23 to 0.79 $m^2\,g^{-1}$ for different trophic states in lakes.

1.6.4 Results and effects

In summary, after considering all the components that absorb radiation, in very clear non-productive oceanic waters blue-green wavelengths in the PAR spectrum dominate, whereas in highly-coloured, humic inland waters blue wavelengths are rapidly attenuated. In humic lakes, CDOM largely governs UV attenuation (e.g. Huovinen et al., 2003; Kirk, 1994a,b; Scully and Lean, 1994)), whereas in oceans (Smith and Baker, 1979) and clear lakes containing low CDOM concentrations the contribution of phytoplankton to UV attenuation can be significant (Sommaruga and Psenner, 1997). If very turbid water contains a large amount of inorganic particles, CDOM is bonded by calcium carbonate contained in the particles, and consequently the colour of the water returns to blue.

There is marked variation in the penetration of UV radiation among water bodies, and within a water body during the year. Global changes, such as climate warming and acidification (Donahue et al., 1998; Schindler et al., 1996; Yan et al., 1996) can lead to increased underwater UV penetration, likewise UV-B radiation itself which can positively affect its own penetration through the photodegradation of CDOM (Morris and Hargreaves, 1997). Variation in the absorption properties of dissolved organic compounds with the seasons and according to their origin and molecular weight (Hessen and Tranvik, 1998; Lean, 1998; Stewart and Wetzel, 1980), interferes with our estimation of UV penetration based on CDOM concentrations. Temporal changes in the absorption characteristics of CDOM have also been reported, with fresh CDOM being photochemically more active than older CDOM (Lean, 1998). It is also notable that UV radiation has been shown to penetrate deeper in saline prairie lakes than in fresh waters of corresponding CDOM concentrations (Arts et al., 2000).

Estimations of CDOM are relatively easy to perform and therefore often used for *in situ* and remote sensing measurements of optical properties. However, another parameter called dissolved organic carbon (DOC, in units of mg l^{-1}) is also useful to measure, since it is more interpretable for studies of carbon cycling and in the context of global change research. Kowalczuk et al. (2010), report that CDOM contributes approximately 20% to the total DOC pool in the open ocean and up to 70% in coastal areas. Unfortunately, on a global scale it has not yet been possible to make a direct link between CDOM and DOC due to the heterogeneous organic composition of CDOM. Until that connection is made, estimation of the universal bulk carbon-specific CDOM absorption coefficient, defined as the ratio of CDOM absorption to DOC concen-

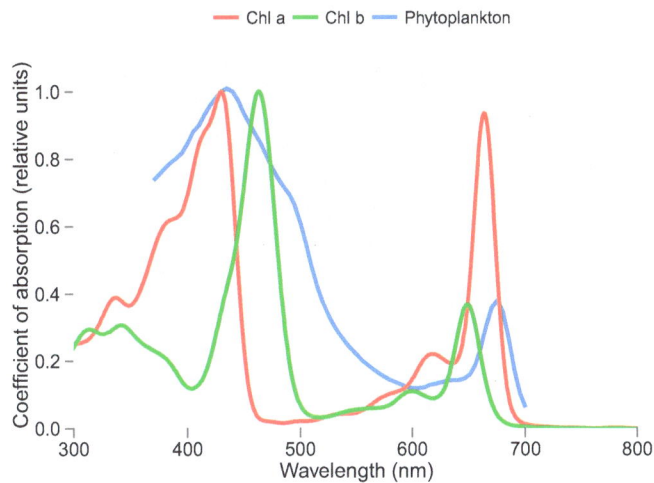

Figure 1.14: Specific oceanic chlorophyll absorption spectrum, after Morel (1991), normalized to maximum absorption at 440 nm (blue line), and absorption spectra of chlorophyll-*a* (normalized at 428 nm, red line) and chlorophyll-*b* (normalized at 436 nm, green line) dissolved in diethyl ether, after Frigaard et al. (1996).

tration, remains almost impossible (Wozniak and Dera, 2007), but at least there are good correlations between a_Y and DOC concentrations in coastal areas (Kowalczuk et al., 2010).

Aquatic organisms can be affected not only directly but also indirectly through UV-dependent changes in the surrounding water, e.g. through increased formation of photochemical reaction products such as singlet oxygen and hydrogen peroxide. Especially in lakes with low DOC concentrations, photoenhanced toxicity of some environmental contaminants or release of complexed metals into the water can occur due to the photodegradation of organic matter (Arfsten et al., 1996; Hessen and Donk, 1994; Palenik et al., 1991; Scully et al., 1997; Zepp, 1982). Despite the potential for detrimental effects, the final impact of UV radiation on organisms may be mitigated by their protective and repair mechanisms (Karentz et al., 1991; Mitchell and Karentz, 1993; Vincent and Roy, 1993), which somehow also depend on certain wavelengths of irradiation. When evaluating the exposure of seaweeds to UV radiation, it should be taken into account that other factors, such as kelp canopies, can markedly reduce the PAR and UV radiation reaching their understorey. Furthermore, the underwater radiation received by seaweeds can be significantly altered depending on the tidal range (Huovinen and Gómez, 2011). Phenolic compounds released from large brown algae into the surrounding water can also locally attenuate UV radiation.

Classifications of water bodies based on their optical characteristics have been developed as general tools in or-

der to allow comparisons between different water bodies. Jerlov (1976) traditional and widely-used classification of marine waters, based on their transmittance of irradiance at different wavelengths, recognizes three oceanic (I–III) and nine coastal (1–9) types of water body (Figure 1.15). Morel and Prieur (1977) classified ocean waters into two types based on their optically dominant components: (i) phytoplankton and their products dominate case-i waters, (ii) particles and dissolved coloured material dominate case-ii waters. The classification proposed by Kirk (1980) is principally suited to inland waters and is based on the spectral absorption of the soluble and particulate fractions. Kirk (1980) defined type G waters, in which CDOM is the dominant light-absorbing component, compared with type T, W and A waters, where tripton, water itself and phytoplankton dominate respectively. Beyond these scales, various other optical classifications have also been proposed (reviewed by Hargreaves, 2003; Kirk, 1994a).

1.6.5 Modelling of underwater radiation

Following Beer's law (Equation 1.8), for deep water (no reflection from bottom), radiance $L(\lambda)$ decreases exponentially with depth z in the water column:

$$L(z, \lambda) = L(z = 0, \lambda) \cdot e^{-\frac{a \cdot z}{\cos \theta}} \quad (1.22)$$

alternatively written as

$$\cos \theta \, \frac{\mathrm{d}L(z, \lambda)}{\mathrm{d}z} = -a \cdot L(z, \lambda) \quad (1.23)$$

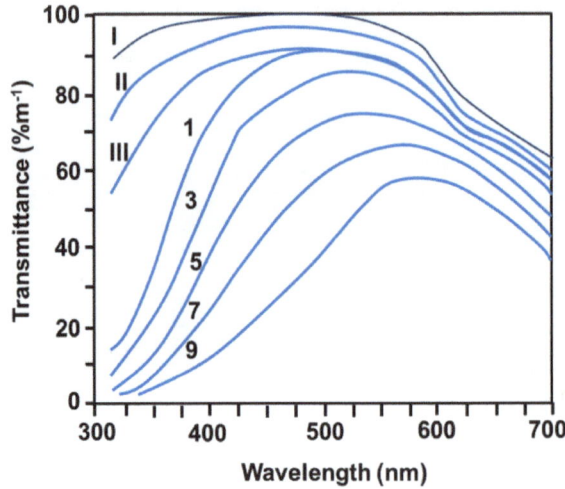

Figure 1.15: Optical classification of marine waters (based on transmittance of irradiance) according to Jerlov's oceanic (I–III) and coastal (1–9) classification system (redrawn from Jerlov, 1976).

θ is the zenith angle of the incoming (downwelling) radiance in water. Note that Eqs. 1.22 and 1.23 are only valid if the water column is homogeneous with depth and there are no scattering particles and internal sources of light in the water, i.e. no fluorescence, raman scattering, nor bioluminescence. Thus, calculation of actual radiative transfer in water (not under idealised conditions) is much more complicated and can only be solved approximately by using empirical, semi-analytical or computational models (Albert and Gege, 2006; Dekker, 1993; Lee et al., 2002; Mobley, 1994).

The radiative transfer equations 1.22 and 1.23 are valid for radiance L, which represents the collimated beam from one specific direction. Due to their construction, radiance detectors do not measure a beam from an infinitely small solid angle, they have an aperture of typically one or two degrees. Other types of detectors sense light from more than one direction, they measure the entire sphere or hemisphere (for details see also section 3.1 on page 71), by integrating the incoming radiance over all directions. Another useful relationship between irradiance E and fluence rate E_0 can be obtained using the Gershun equation:

$$\frac{\mathrm{d}}{\mathrm{d}z} E(z, \lambda) = -a\, E_0(z, \lambda) \qquad (1.24)$$

If, for example, only the downwelling irradiance $E_d(z, \lambda)$ is measured or necessary for calculating radiative transfer, Equation 1.23 yields

$$\frac{\mathrm{d}E_d(z, \lambda)}{\mathrm{d}z} = -K_d \cdot E_d(z, \lambda) \qquad (1.25)$$

giving the diffuse attenuation coefficient for downwelling irradiance, K_d. K_d is related to the total absorption a and

scattering b as well as to θ and also the solar zenith angle θ_s (Albert and Mobley, 2003; Kirk, 1991). A practical and often-used method to estimate K_d, and therefore the transparency of the water, is the Secchi disk test. The visibility of a submerged white disk can be correlated to downwelling diffuse attenuation (Tyler, 1968). This and also the penetration depth z_d give useful information about the water body. The penetration of irradiance important for photosynthesis (primary production) is often expressed as the depth at which, for example, 1% or 10% of the value just below the water surface is reached. The depth, where 1% of PAR is reached, separates the euphotic zone from the aphotic zone. After Kirk (1994a), z_d can be obtained from K_d: $z_d(1\%) = 4.6/K_d$ and $z_d(10\%) = 2.3/K_d$. In Figure 1.16, an example of spectral attenuation of solar radiation at different depths, as well as the penetration depths of UV-B, UV-A radiation and PAR are given for coastal waters of the south-eastern Pacific Ocean (off the coast of Chile).

A very useful tool to simulate spectral radiance and irradiance under water, depending on different concentrations of the water constituents or bottom depth and type, is called WASI (water colour simulator, Gege, 2004). The software includes different analytical parameterisations and it can also be used for inverse calculations, i.e. for estimating optical properties and concentrations of water constituents from (remote sensing) measurements. The software program including the manual is available free of charge using an anonymous login at the ftp server `ftp://ftp.dfd.dlr.de/pub/WASI`. Other models, which include angular distributions of radiation under water are, for example, the Monte Carlo method (Prahl et al., 1989; Wang et al., 1995)[10], HydroLight (technique described by Mobley, 1994), or EcoLight-S (Mobley, 2011).

1.7 UV radiation within plant leaves

Modelling of UV-B within plant leaves has not so far been successfully achieved. For empirical estimations of UV penetration two methods have mainly been used: fibre-optic measurements and UV-induced fluorescence. Neither of them is ideal and both give only partial information.

1.7.1 Fibre-optic measurements

This method was introduced for visible light by Vogelmann and Björn (1984), and has been adapted for ultraviolet radiation by Vogelmann, Bornman and coworkers (Bornman and Vogelmann, 1988; Cen and Bornman, 1993; DeLucia et al., 1992). The method cannot be used for absolute measurements due to uncertainties about the local conditions and the acceptance angle at the fibre tip, but it has yielded valuable comparisons between wavelengths and depth distributions of radiation. Fibre probes can be made more angle-independent (García-Pichel, 1995), but then become too bulky for measurements inside plant leaves.

1.7.2 UV-induced chlorophyll fluorescence

This method has so far only been employed by excitation and measurement at the leaf surface. It thus mainly monitors penetration of UV radiation through the epidermis into the chlorophyll-containing mesophyll, i.e. transmission through the epidermis (Figure 1.17). Chlorophyll fluorescence excited by blue light usually serves as a standard (Bilger et al., 2001, 1997), more recently excitation by red light has also been used to avoid interference by anthocyanins (Goulas et al., 2004). The principle has been used in commercial instruments that assess the UV absorbance of the epidermis (Goulas et al., 2004; Kolb et al., 2005). The first portable instruments, *UVA-PAM* and *Dualex FLAV* used UV-A radiation for excitation rather than UV-B radiation, but now there is at least one instrument, *Dualex HCA*, measuring UV-B absorbance.

In principle, chlorophyll fluorescence could also be used for monitoring UV penetration in another way, by recording fluorescence in cross-sections of leaves, as has already been done for the penetration of visible light (Gould et al., 2002; Vogelmann and Evans, 2002; Vogelmann and Han, 2000).

1.7.3 Factors affecting internal UV levels

Many studies have indicated that UV-absorbing compounds in the vacuoles of epidermal cells have a major role in regulating internal UV levels. However, absorbing compounds located in cell walls and other cell parts can also be important in controlling internal UV penetration. These compounds are usually not easily extractable, and consequently one cannot rely on extracts alone when judging the effectiveness of UV-screening protection. Furthermore, both wax deposits and pubescence may be very important for protection against ultraviolet radiation (Holmes and Keiller, 2002; Karabourniotis and Bornman, 1999). The amount of UV protection can also vary over time, even during the day, probably as a result of changes in flavonoid concentrations (Barnes et al., 2008; Veit et al., 1996). Of course leaf properties conferring protection also depend on environmental factors, in particular prior UV-B exposure, and great differences exist among plant species.

1.8 Action and response spectra

Plants do not respond equally to all wavelengths of UV, and this spectral response can be described by a response spectrum and/or by an action spectrum. It is important to be aware that an action spectrum is not the same thing as a response spectrum. Although they are both used to describe the wavelength dependency of a biological response to radiation they are measured and used differently. Because they are measured in different ways they yield curves of different shapes. A response spectrum shows the size of the response at a *fixed photon fluence*[11] of radiation across a series of different wavelengths. However, since UV radiation never comes at fixed irradiances over the entire spectrum the response spectrum is of limited use when estimating the physiological response to solar- or broadband UV exposure. In contrast, an action spectrum shows the effectiveness of radiation of different wavelengths in *achieving a given size of response*. This is a very important difference because dose response curves are not necessarily parallel or linear.

When estimating biologically effective irradiances (see section 3.10 on page 99) it is very important to use appropriate action spectra for each biological process. However, since the action spectra of many biological responses are not known, this provides a dilemma for researchers who must try to chose the action spectrum that best approximates the process they are studying. Using the wrong action spectrum can produce very large errors

---

[10]Program available from the website of the Oregon Medical Laser Center at http://omlc.ogi.edu/software/mc/

[11]Different values of photon fluence can be obtained either by varying the irradiance, or the irradiation time. However, if the irradiation time is varied it is important to check that reciprocity holds. In other words, that the same fluence achieved through different irradiation times elicits the same size of response.

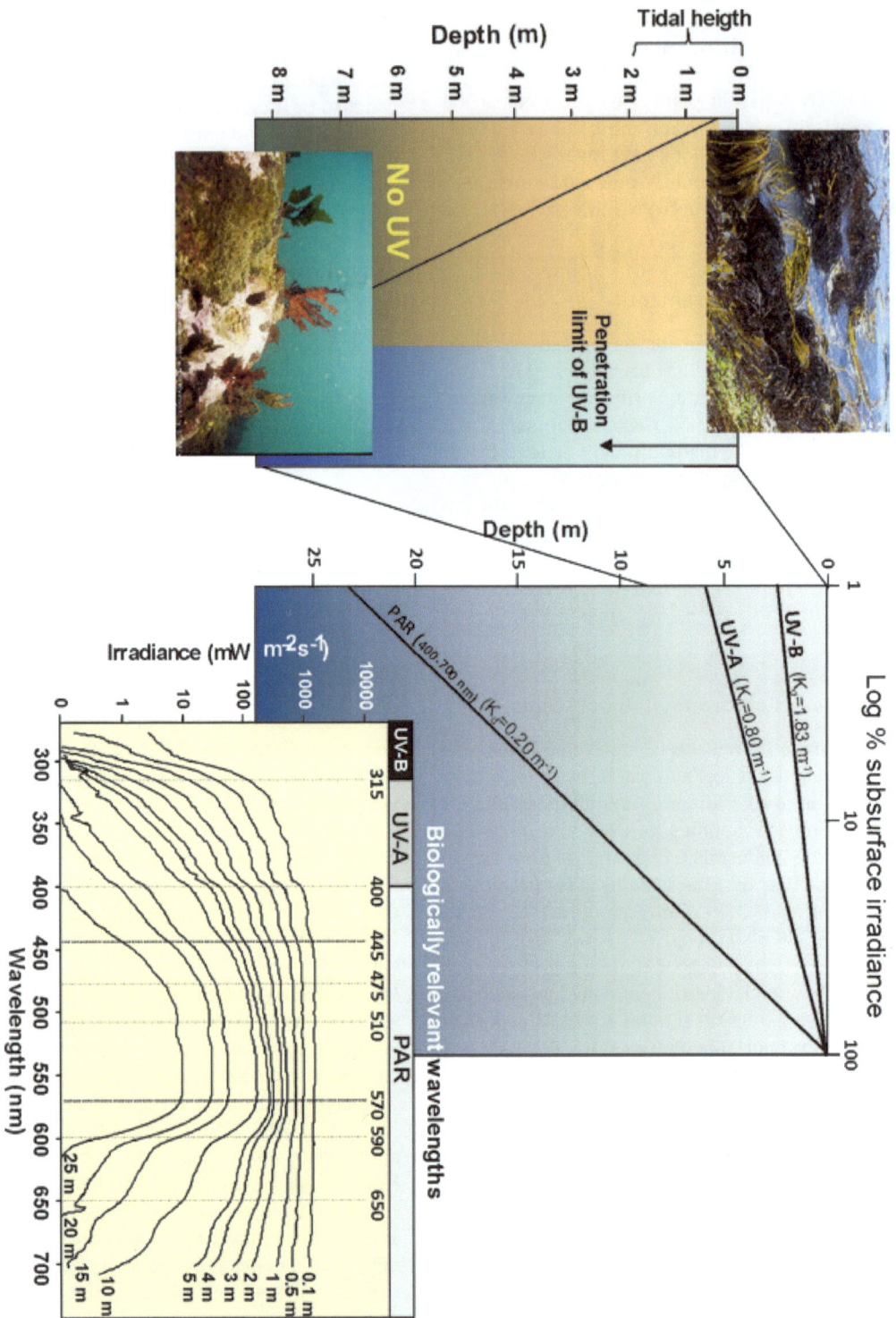

Figure 1.16: Example of the underwater irradiance in a coastal system in the Pacific Ocean off southern Chile (coast of Valdivia).

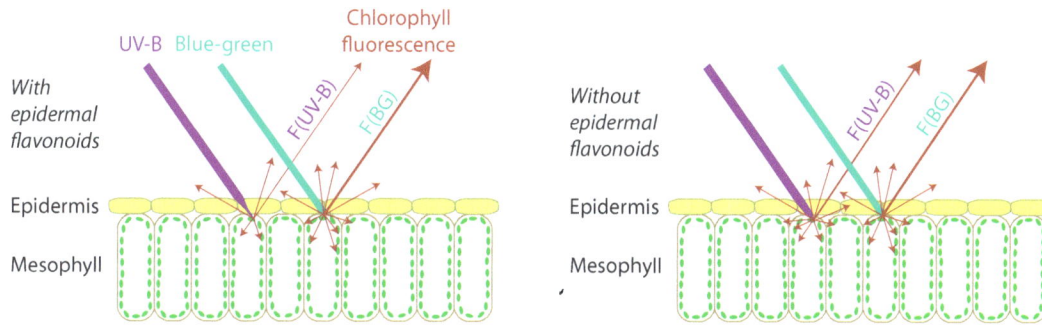

Figure 1.17: Chlorophyll fluorescence and UV radiation penetration into leaves. The reference excitation light, blue-green in this example, is not absorbed by the epidermis, while the UV radiation is partly absorbed. Fluorescence produced by UV-radiation excitation, F(UV), and by blue-green light excitation, F(BG), can be used to estimate UV absorbance by the epidermis. Redrawn and adapted from Bilger et al. (2001).

when predicting effective irradiance and, therefore, UV effects on plants due to e.g. ozone depletion (Cullen and Neale, 1997; Rundel, 1983). This is also a good reason for continued efforts to measure action spectra.

Building a response spectrum is fairly simple; we need to measure a plant response at a single (photon) irradiance (or fluence) for each wavelength (or narrow band), whereas for constructing an action spectrum ones needs to measure the response at several different photon fluence values for each wavelength of interest.

1.8.1 Constructing a monochromatic action spectrum

Action spectra are most frequently measured using monochromatic light, i.e. radiation of a single or narrow range of wavelengths. This can be achieved by the use of systems which transmit, or emit, only a defined and usually narrow range of wavelengths, e.g. band-pass filters, LEDs, spectrographs, or lasers (see Chapter 2 for details about radiation sources). It is also possible to build, for example, an UV action spectrum with background irradiation of other wavelengths such us PAR.

Shropshire (1972) describes in detail the theory behind monochromatic action spectra, and the assumptions needed for an action spectrum to match the absorption spectrum of a photoreceptor pigment. He also considers the problem of how screening by other or the same pigments can distort the shape of action spectra. He gives examples for visible radiation but the theoretical considerations are fully applicable to ultraviolet radiation.

To construct a true action spectrum we first need to measure dose response curves for radiation of different wavelengths (Figure 1.18). The more curves we measure and the narrower the wavelength range used for each of these, the more spectral detail will be visible in the action spectrum built from them. For each of these curves, we should use a range of photon fluences yielding response sizes going from relatively small responses to close to the maximum response size (close to saturation). The photon fluence values used should increase exponentially. Photon fluence can be varied both by varying irradiance and/or irradiation times. If irradiation time is varied, it should be checked that reciprocity holds[12]. We fit a curve to each set of dose response data, using the logarithm of the photon fluence as an independent variable. Using a logarithmic scale is expected to yield a more linear response curve than untransformed photon fluence values. From the fitted dose-response curves we calculate by interpolation the photon fluence required at each measured wavelength to obtain a response of the selected target response size. We use the photon fluence values to calculate effectiveness as $1/(Q \cdot t)$ where fluence is given by the product of photon irradiance (Q) by the irradiation time (t), and we finally plot these effectiveness values against wavelength (λ). If the dose response curves are not parallel, the shape of the action spectrum will depend on the target response level chosen. Different causes have been suggested for the lack of parallelism of dose response curves that is sometimes observed. Two of these suggestions are self-screening effects and involvement of two or more interacting pigments in the response (Shropshire, 1972). One can in principle use either quantum (=photon) or energy units, but quantum units are preferable as for any photochemical reaction absorption events always involve quanta. The shape of the spectrum will depend on whether an energy or photon basis is used.

As the main feature of interest is the shape of the curve, UV action spectra are usually normalized to an

---

[12]Reciprocity refers to the assumption that equal values of photon fluence achieved by irradiation differing in length and photon irradiance, but supplying the same total number of photons, are expected to elicit an identical response.

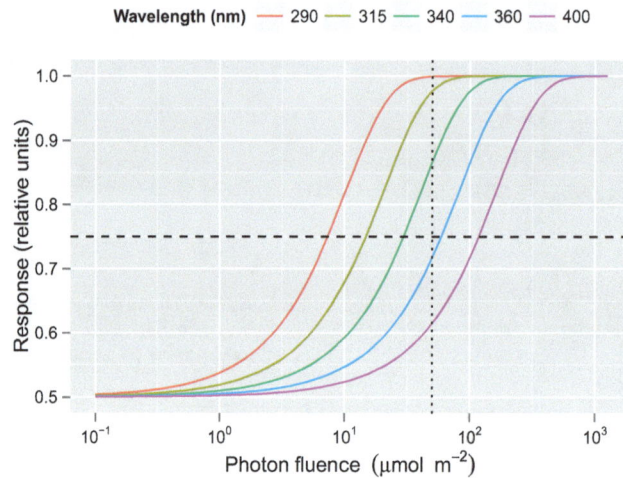

Figure 1.18: Hypothetical example of dose response curves at different wavelengths. The horizontal dashed line defines an action spectrum (equal response size), and the vertical dotted like defines a response spectrum (equal photon fluence). Adapted from Gorton (2010).

action that is equal to one at 300 nm. This is achieved, by dividing all the quantum (or photon) effectiveness values measured at different wavelengths, by the effectiveness at 300 nm. The use of $\lambda = 300$ nm is an arbitrary convention, of rather recent adoption, so you will find, especially in the older literature, other wavelengths used for the normalization.

A response spectrum will rarely match the absorption spectrum of the photoreceptor because of non-linearities in later steps between light absorption by the photoreceptor and an observed response. In the case of action spectra, by keeping the size of the response constant across wavelengths we attempt to minimize the effect of these non-linearities on the measured spectrum. Because of this, a properly measured action spectrum will usually closely follow the absorption spectrum of the pigment acting as photoreceptor, except from possible effect from interfering pigments. Figure 1.19 and Table 1.4 show several action spectra relevant to research on the effects of UV on plants. See the article by Gorton (2010) for a deeper discussion on biological action spectra.

1.8.2 Constructing a polychromatic action spectrum

Monochromatic action spectra are useful to understand the nature of a specific response, e.g. damage, but are not suitable for calculating real effects under solar radiation. The response of a plant to light and UV radiation depends on both the amount of energy (dose-response)

and the spectral composition of the radiation. Polychromatic action spectroscopy is based on a background of broad-band white light from artificial sources or natural daylight supplemented by various wavelength, for example between 280 and 360 nm (Holmes, 1997). This polychromatic approach provides an action spectrum useful for assessing effects of UV under normal plant growing conditions, because the simultaneous exposure to a broad wavelength interval has a different net effect than that of an exposure to separate monochromatic radiation, due to synergisms or antagonisms between complex chemical and biological processes, for example repair mechanisms (Coohill, 1992; Madronich, 1993). In practice, such a realistic polychromatic radiation spectrum is achieved using a series of cut-off filters, which cut off radiation of wavelengths shorter or longer than a certain wavelength. The effect of such filters can be compared to the effect of the variable thickness of the stratospheric ozone layer. Figure 1.20 shows examples of spectral irradiance for broad band light from lamps filtered by different cut-off filters. The cut-off wavelength usually refers to the wavelength of 50% transmission.

The original approach, so called differential polychromatic action spectroscopy, is described by Rundel (1983). Different biological responses are proportional to different specific treatments, for example UV exposures. Thus, an action spectrum can be estimated by quantifying differences in responses between successive treatments. The number of treatments required depends on the relative change in response over a wavelength interval. One

Table 1.4: Some action spectra used in research on the effects of UV radiation on plants, abbreviations and references to original definitions (for details see Kotilainen et al., 2011). See Figure 1.19 for plots for four of these spectra, and Appendix 3.17 on page 111 for mathematical formulations.

Code	Full name	Source
CIE	Erythemal, standardized by CIE	McKinlay and Diffey (1987)
DNA(N)	'Naked' DNA	Setlow (1974)
DNA(P)	Plant DNA, *Medicago sativa*	Quaite et al. (1992)
FLAV	Flavonol (mesembryanthin) accumulation, *Mesembryanthemum crystallinum*	Ibdah et al. (2002)
GEN	Generalized plant action spectrum, composite	Caldwell (1971); Flint and Caldwell (1996)
GEN(G)	Green's formulation of GEN	Green et al. (1974)
GEN(T)	Thimijan's formulation of GEN	Thimijan et al. (1978)
PG	Plant growth, *Avena sativa*	Flint and Caldwell (2003)
PHIN	Photoinhibition of isolated chloroplasts	Jones and Kok (1966)

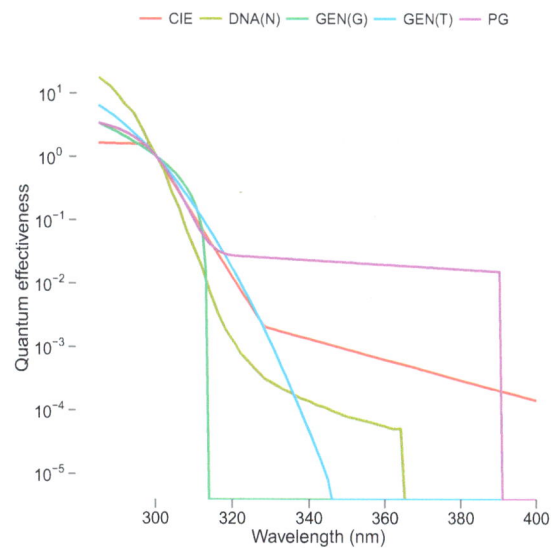

Figure 1.19: Some action spectra used in research on the effects of UV radiation on plants. The spectra have been normalized to action equal to one at a wavelength of 300 nm. See Table 1.4 for codes used for the spectra and Appendix 3.17 for the formulations.

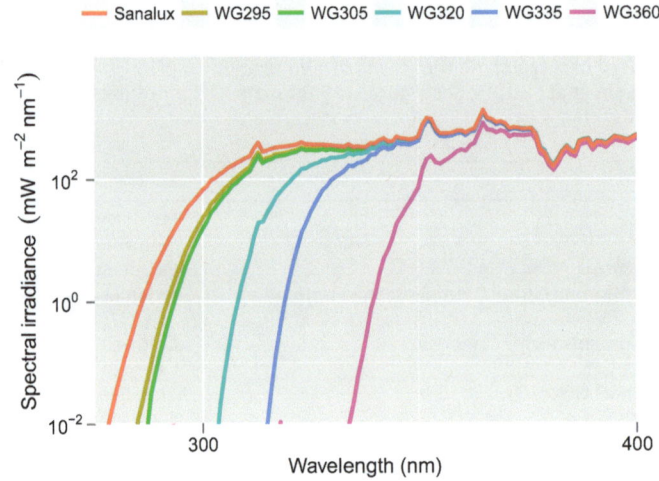

Figure 1.20: Spectra of the different UV scenarios used in an experiment. Numbers refer to the UV cut-off position (50 % transmission) produced by different Schott coloured glass filters (thickness 3 mm, Schott, Mainz, Germany): WG295, WG305, WG320, WG335, and WG360 in combination with a layer of Sanalux glass (thickness 4 mm, Schott).

example of a quantifiable effect would be change in the observed concentrations of flavonols.

1.8.2.1 The mathematics behind polychromatic action spectra

The average proportionality $\bar{s}$ between differences in the biological effective response of successive treatments i, $\Delta W_{be,i}$, and differences in exposure (fluence rate multiplied by exposure time), ΔH_i is given by an average quantity

$$\bar{s_i} = \frac{\Delta W_{be,i}}{\Delta H_i} \qquad (1.26)$$

All the values of $\bar{s_i}$ together are represented in the action spectrum. If the wavebands of each treatment are small enough, an action spectrum $s(\lambda)$ can be expressed by a mathematical function. Different functions are discussed (Cullen and Neale, 1997; Rundel, 1983). In the simplest case the factors $\bar{s_i}$ are positive and decrease exponentially with increasing wavelength. Thus, one common form of $s(\lambda)$ is

$$s(\lambda) = e^{-k(\lambda - \lambda_0)} \qquad (1.27)$$

where k is a parameter, which has to be obtained for each different biological effect using a fitting procedure, and a wavelength λ_0, where the action spectra are normalised to unity, e.g. $\lambda_0 = 300$ nm. Other functions including a polynomial dependence on λ are also possible and perhaps necessary for describing complex mechanisms.

For the interpretation of the experimental data, the biological effective response W_{be}, it is necessary to consider the wavelength dependency. Thus, it is crucial that the entire spectral irradiance $E(\lambda, t)$ during the experiment is known. Broad-band meters are not suitable in most cases. This dependency is included in the biological effective dose function (exposure) H_{be}, given by

$$
\begin{aligned}
H_{be} &= \int_\lambda \int_t s(\lambda) \cdot E(\lambda, t) \ \mathrm{d}t \ \mathrm{d}\lambda \\
&= \int_\lambda s(\lambda) \cdot H(\lambda) \ \mathrm{d}\lambda \qquad (1.28)
\end{aligned}
$$

A mathematical model can separate wavelength and dose dependency for a set of experimental data by the use of mathematical functions describing $W_{be}(H_{be})$ and optimisation procedures, e.g. non-linear curve fitting (Cullen and Neale, 1997; Ghetti et al., 1999; Götz et al., 2010; Ibdah et al., 2002). The model assumes that photons at different wavelengths act independently, but with different quantum efficiency at the same absorption site, and therefore with the same mechanism. Regarding the shape and saturation of the observed UV effect, different functions can describe the data of $W_{be}(H_{be})$, for example a linear, hyperbolic or sigmoid function. A simple linear relation is

$$W_{be}(H_{be}) = W_0 \cdot H_{be} \qquad (1.29)$$

with the parameter W_0, which has to be determined by the fitting procedure. If for example the exposure time of the UV radiation is constant and, therefore, not depending on the duration of the experiment t, the combination of Eqs. 1.27 to 1.29 yields

$$W_{be}(H_{be}) = W_0 \cdot t \cdot \int_{\lambda} e^{-k(\lambda-\lambda_0)} \cdot E(\lambda) \ d\lambda \qquad (1.30)$$

To solve the equation for the unknown parameters W_0 and k, measurements of the biological response W_{be} and the spectral irradiance $E(\lambda, t)$ have to be put into a fitting routine, to optimise W_0 and k by minimising the differences, e.g. the (root) mean square, between measured and modelled values of W_{be}. In Figure 1.20 are six different UV scenarios shown. In this case, the differences among six individual measured and modelled responses have to be calculated and the sum of these needs to be minimised by a non-linear optimisation technique, such as that provided by the add-in "Solver" in Excel.

1.8.3 Action spectra in the field

Under field conditions it is more difficult to build UV action spectra, and most frequently what are measured are polychromatic action spectra (e.g. Cooley et al., 2000; Keiller et al., 2003).

As many whole-plant responses result from a long signal transduction chain, which depends on the action of more than one photoreceptor, the action spectrum for many responses at the whole-plant or organ level, seems to vary among species, with the seasons of the year or growing conditions. For responses like these it is almost impossible to define a unique and stable action spectrum for plants growing outdoors, as these responses are too far decoupled from the photoreceptors.

Even if they do not faithfully reflect the properties of a single photoreceptor, action spectra can be extremely useful, as we need them as biological spectral weighting functions (BSWFs) when calculating biologically effective UV doses (see section 3.10 on page 99), since the same UV radiation spectrum has a different effect on different plant responses (Figure 1.19).

1.9 Further reading

http://www.photobiology.info/, photobiological sciences online. At this web site there are many articles, several of them relevant to plant photobiology. The book *Photobiology: The Science of Life and Light* by Björn (2007) is a general introduction to photobiology, that complements well this chapter. The mechanism of ozone depletion and its consequences has been accessibly described by Graedel and Crutzen (1993) in the book *Atmospheric Change: An Earth System Perspective*. The UNEP reports (UNEP, 2003, 2007, 2011, and earlier) provide up-to-date reviews on the environmental consequences of stratospheric ozone depletion.

1.10 Appendix: Calculation of polychromatic action spectra with Excel (using add-in "Solver")

This is an example of one possibility to derive the parameters of a polychromatic action spectrum by non-linear optimisation as explained in section 1.8.2 on page 24. The Excel add-in "Solver" is listed in the menu "Tools". If this is not the case, it has to be installed in Tools>Add-Ins. This experiment was performed in the small sun simulator of the Helmholtz Zentrum München, Neuherberg, Germany (see section 2.2.7 at page 48) in a special cuvette, which allows simultaneous exposure of plants under different UV radiation (Götz et al., 2010; Ibdah et al., 2002). Figure 1.21 shows the cuvette placed in the sun simulator with five rows of glass filters from Schott, Mainz, Germany (WG295, WG305, WG320, WG335, WG360).

At the beginning we have to put all our measured data into different sheets in Excel. In this example, the biological response W_{be} is the UV-induced flavonoid lutonarin (in μmol g^{-1} FW (fresh weight)) of the first leaf of the barley cultivar "Barke". The amount was estimated by HPLC (high performance liquid chromatography). Three young plants in each of three independent experiments were harvested under six different UV scenarios, which spectral irradiances are shown in figure 1.20 in section 1.8.2 on page 24. The data are put into the first sheet "leaf data" and mean values and standard deviations for each UV scenario were calculated as illustrated in figure 1.22.

The second sheet "spectra" contains all measured spectra of the six different UV scenarios under the glass filters. For the calculation of a polychromatic action spectrum in the UV range it is recommended that the spectrum is measured using a double-monochromator system from 280 to 400 nm in steps of 1 nm. In this experiment, the irradiance was increased from the morning until noon and then decreased until the evening, to simulate natural variation in solar radiation. This was done in four steps as shown in the picture of the third sheet "daily exposure" in figure 1.23. As an example, typical radiation data under the glass filter WG305, measured during the experiment, are presented in table 1.5. This sheet is only necessary to calculate the exposure time at each light level in seconds as shown in the marked cell F8. Sheet "spectra" contains the wavelengths in column A and all the 24 spectra (six UV scenarios and four light levels) from column B to Y.

Now all sheets for input data are finished and the sheets for solving our system of equations has to be prepared. Therefore, the next sheet "weighted spectra" is filled with the information about the action spectrum

Figure 1.21: Special cuvette in the small sun simulator of the Helmholtz Zentrum München (see section 2.2.7 at page 48) covered by five coloured glass filter from Schott, Mainz, Germany, to simulate five different UV scenarios. The filters from top to bottom are WG295, WG305, WG320, WG335 as well as WG360. The resulting spectral irradiances in the UV range for one specific incoming radiation is shown in figure 1.20 in section 1.8.2 on page 24. Photograph: Andreas Albert.

Table 1.5: PAR and UV radiation during the experiment for barley under the coloured glass filter WG305 (Schott, Mainz, Germany).

Light level	1	2	3	4	Unit
$E_{\text{UV-B}}$	0.152	0.584	1.135	1.531	W m^{-2}
$E_{\text{UV-A}}$	8.76	19.78	24.67	38.74	W m^{-2}
E_{PAR}	114	228	256	363	W m^{-2}
PPFD or Q_{PAR}	537	1073	1198	1692	μmol m^{-2} s^{-1}

function as shown in figure 1.24. This sheet contains the calculation of the biological effective dose function H_{be} as presented in equation 1.28 on page 26. First, for the action spectrum $s(\lambda)$ an exponential function in chosen, as explained for equation 1.27 on page 26. Column A includes the wavelengths and column B the estimated value of the action spectrum using the parameter $\lambda_0 = 300$ nm of cell B4 and $k = 0.200$ nm^{-1} of cell B5 (marked red in figure 1.24). This cell is only a link to cell B11 in the next sheet "SOLVER". Because it is only the starting value of the optimisation procedure, the values of column B do not represent the real action spectrum. These values will change during optimisation. Below row 6, column D and the following columns (in this case to column AA) include the values of the action spectrum of column B multiplied by the spectral irradiances of each UV scenario and light levels of sheet "spectra". Row 1 and 2 include the estimation of the UV doses. The cells in row

2 from column D and the following columns represent the integration over the wavelengths from 280 to 400 nm of each column beneath multiplied by the exposure time of the respective UV scenario and light level as previously calculated in the sheet "daily exposure". Integration in Excel is done by adding the values of the cells regarding the wavelength step as shown for the marked cell D2 in figure 1.24. Here it is 1 nm. To get doses in units of Ws m^{-2} (J m^{-2}), the sum is divided by a factor of 1000 because the spectral irradiances were measured in mW m^{-2} nm^{-1}. The daily UV dose (in kJ m^{-2}) is then calculated in row 1 for each UV scenario by adding the four values of the light levels. These daily UV doses are now needed for further calculations in the next sheet "SOLVER".

The sheet "SOLVER" as illustrated in figure 1.25 includes the comparison of the measured and modelled data—as table and figure. The first five rows are a short explanation of the entire model. Especially row 5 con-

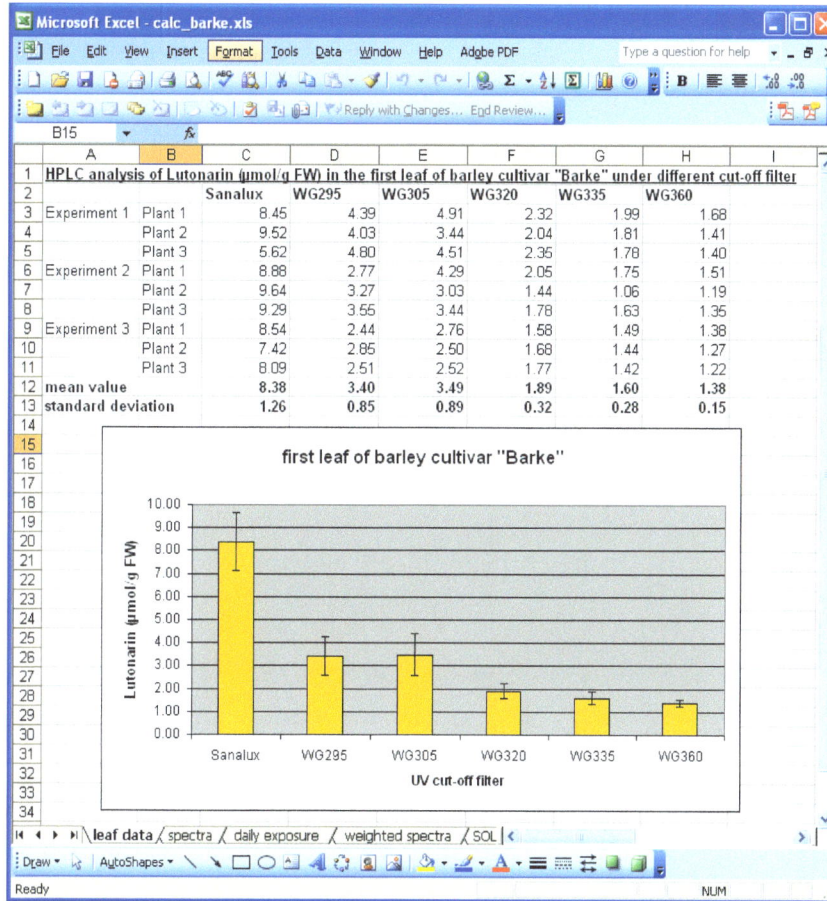

Figure 1.22: Non-linear optimisation for estimating polychromatic action spectra: Excel sheet including input data of the UV-induced flavonoid lutonarin in the first leaf of young plants of barley cultivar "Barke" obtained by HPLC (high performance liquid chromatography).

tains the mathematical function describing the measured biological response W_{be}, which was the flavonoid content of lutonarin in the first leaf of barley. Here, a sigmoid function was chosen, given by

$$W_{be} = \frac{W_0}{1 + e^{-\frac{H_{be} - H_0}{b}}} \qquad (1.31)$$

Using this function, the values of each UV scenario of row 15 were calculated. Therefore, the variable coefficients W_0, H_0, b, and k as well as their constraints from row 8 to 11 were used. Row 14 includes the mean values of each treatment derived in the first sheet "leaf data". The squared differences of measured and modelled data are calculated in row 16 and added up in cell C17. All cells are now prepared and we can start to optimise our variable coefficients by minimising the value of cell C17. Thus, we mark this cell by clicking on it and choose from the menu `Tools>Solver`. A new small window called "Solver Parameters" will appear as shown in the left part

of figure 1.26. Here, the target cell was C17 and the box for minimisation was checked. The optimisation was done by changing the cells B8 to B11 concerning all constraints listed in the sheet. By clicking on the button "Options", another window will pop up, where the parameters for the optimisation algorithm are defined (right part of figure 1.26). In this example, a non-linear Newton method was used. The parameters of the upper left part of the window define the number of iterations or the maximum time for calculation as well as the precision or tolerance of the result. After inserting the numbers and checking the necessary boxes, click "OK" to return to the window "Solver Parameters". Here, the optimisation starts by clicking on "Solve". After a while, the program returns a message if the search succeeded and by accepting, the result is written into the cells of sheet "SOLVER" chosen for changing.

The final result for the data in this example is shown

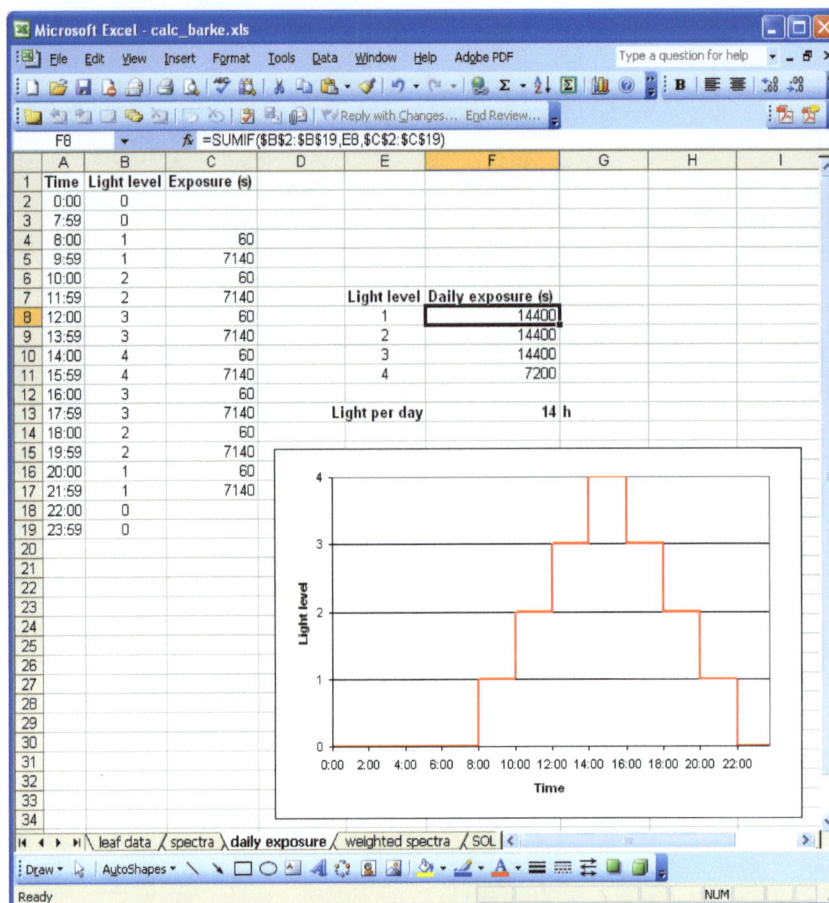

Figure 1.23: Non-linear optimisation for estimating polychromatic action spectra: Excel sheet including the information about the daily variation of light levels and the derivation of exposure time of each light level.

in figure 1.27. The sum of the squared differences was minimised to 0.13. The left graph in the bottom of figure 1.27 shows two bars for each UV scenario, one for the measured leaf data of row 14 (yellow) and the other one for the modelled data of row 15 (purple). The result looks promising. The most important parameter regarding the action spectrum is $k = 0.126$ nm^{-1} in cell B11. This value was used to plot the actual action spectrum of this study in the upper right graph of figure 1.27. Our action spectrum is valid for wavelengths between 280 and 400 nm,

that interval which was chosen for integration in sheet "weighted spectra".

This example describes just one possibility to solve a multi-variable and non-linear problem. For example, there is also a plugin available for Calc in OpenOffice.org. Therefore, if using Calc instead of Excel, the file NLPSolver.oxt[13] has to be installed by the extension manager. Other search algorithms are used in this tool, but the result will be the same.

---

[13]available from http://extensions.services.openoffice.org/project/NLPSolver.

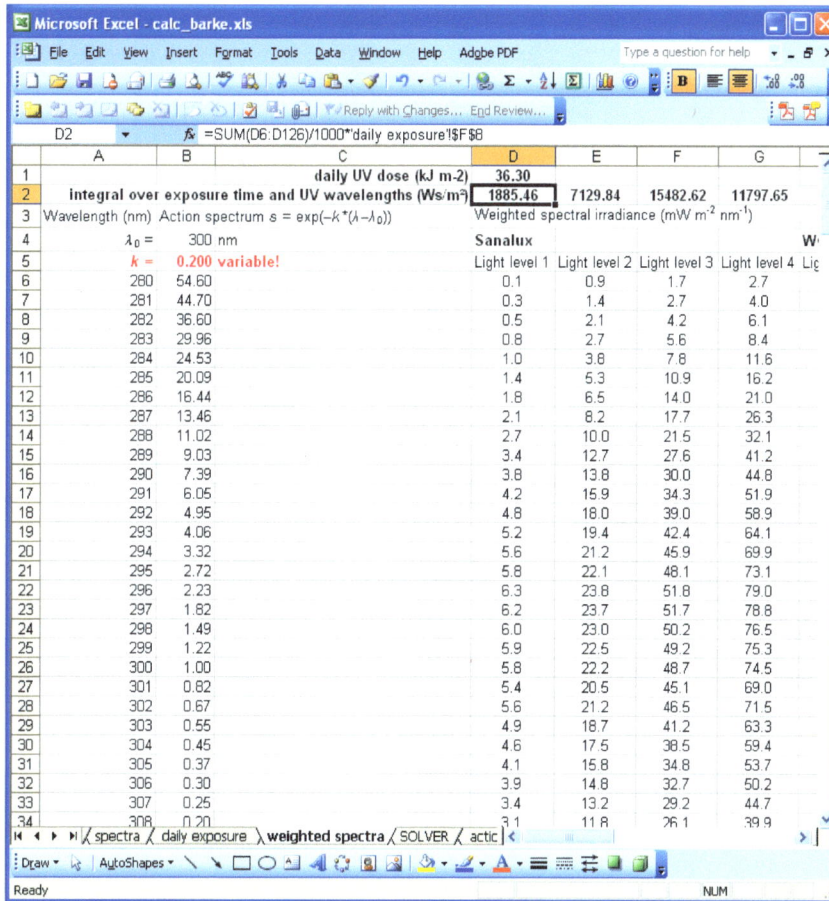

Figure 1.24: Non-linear optimisation for estimating polychromatic action spectra: Excel sheet including the calculation of biological effective dose function by integrating spectral irradiances weighted by the action spectrum over UV wavelength and exposure time.

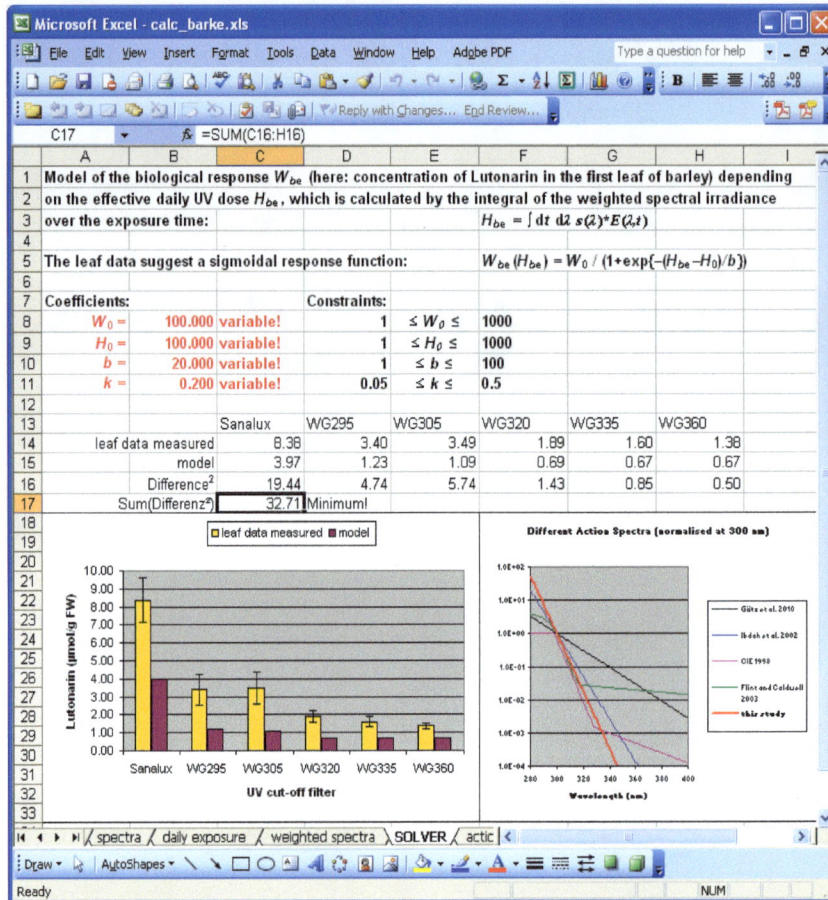

Figure 1.25: Non-linear optimisation for estimating polychromatic action spectra: Excel sheet with starting values of all variable coefficients for the model, as table and graph, and the sum of squared differences between modelled and measured flavonoid content of lutonarin, which has to be minimised. The right graph at the bottom includes the actual action spectrum of this study compared to other action spectra from the literature.

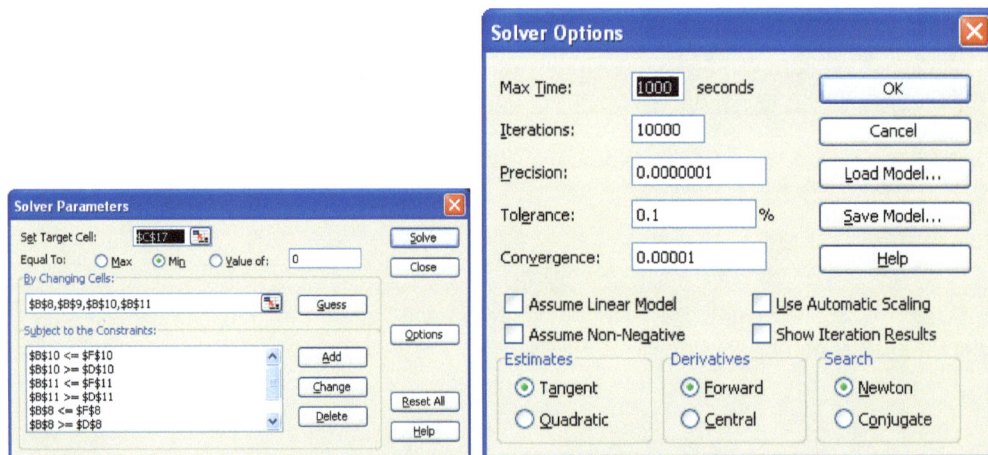

Figure 1.26: Popup windows for defining parameters (left) and options (right) in the Excel add-in "Solver".

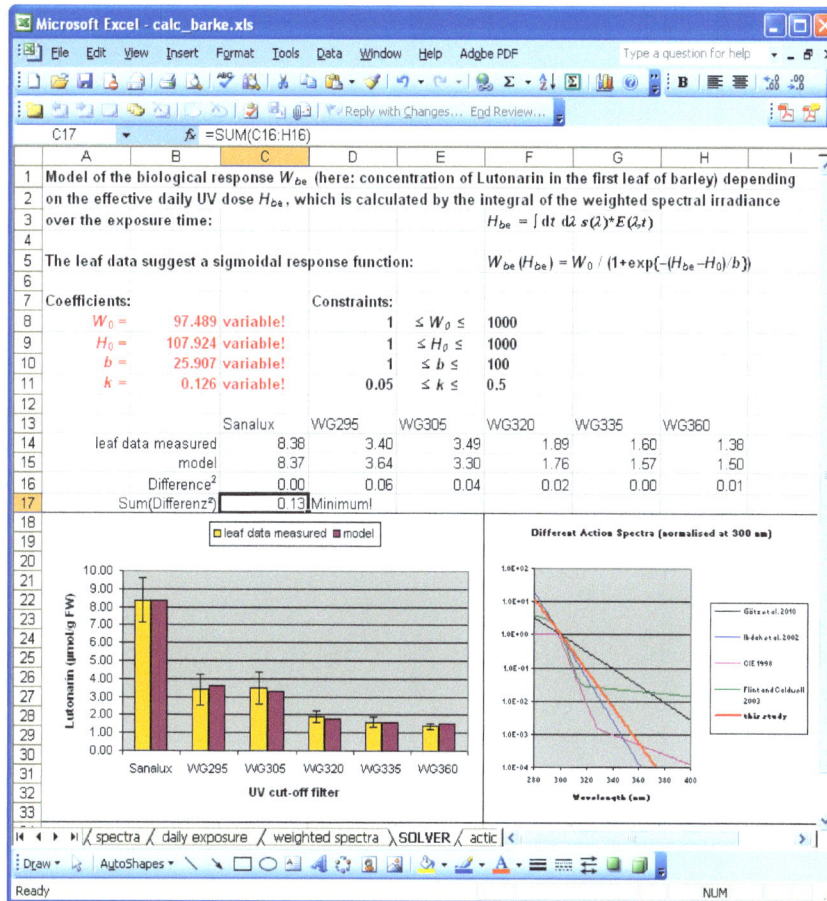

Figure 1.27: Non-linear optimisation for estimating polychromatic action spectra: excel sheet with the results for all variable coefficients after running the search algorithm.

2 Manipulating UV radiation

Pedro J. Aphalo, Andreas Albert, Andy McLeod, Anu Heikkilä, Iván Gómez, Félix López Figueroa, T. Matthew Robson, Åke Strid

2.1 Safety considerations

2.1.1 Risks related to sunlight exposure

When working in field experiments, workers and researchers are exposed to sunlight as in any other outdoor activity. To avoid health hazards adequate protection should be used, either in the form of clothing or sun-blocking lotions. In some cases, for example when there is strong reflection of UV radiation by water, sand or snow-covered surfaces, eye protection in the form of sunglasses or goggles should be used.

A common and simple parameter of the UV radiation level outdoors is the UV index, which was developed by the World Health Organization (WHO) including recommended protection (WHO, 2002). The UV index is used for example in weather forecasts in integer numbers from 1 to 11+, representing the UV exposure from low to extreme. [1]

2.1.2 Risks related to the use of UV lamps

Lamps emitting UV-B and UV-C create a higher risk of eye damage than sunlight because when UV radiation is not accompanied by strong visible light the eye pupils remain wide open. Also many lamps emit UV-C, which is not present in sunlight at ground level and is more damaging than UV-B. It is important to use goggles to protect the eyes not only from UV radiation from the front but also from the sides of the face. It is important to ensure that the goggles used are designed to protect from UV radiation, rather than just from the impact of flying particles. The skin should also be well protected, because in the same way as for plants, the effect of UV-B on humans is enhanced under low visible light irradiance by the low rate of photoreactivation (light-driven repair of DNA damage).

Locations where UV lamps are in use should have visibly located warning signs. Although there is no standardized symbol for UV radiation hazard, the warning symbol for optical radiation (Figure 2.1) accompanied by the text 'UV radiation' can be used. Additional text, 'protect eyes and skin' or 'do not look directly at light source' can be added. Access to outdoor experiments should be restricted by fences with locked gates, and by locking access doors to greenhouses, controlled environment rooms and cabinets. This is to prevent exposure of people who are unaware of the risks involved. Everybody with access to the area should be informed of risks, and trained to use the protection and work procedures required to mitigate them.

Lamps also pose risks unrelated to UV radiation. Fluorescent tubes contain mercury and should disposed of as hazardous waste. The glass envelope of lamps is fragile, and can cause injuries if it breaks. Xenon-arc lamps also present a relatively high risk of explosion, goggles and other face and body protection should be used. This risk is caused by the high pressure inside this type of lamp and is present irrespective of the xenon-arc lamps being on or off.

Lasers, can easily produce injuries, and should be handled with great care. Even the very low power laser pointers used during lectures are capable of injuring the eye if pointed towards the audience. As with UV light sources, access to any type of laser should be restricted to trained users, and eye protection should be worn to

---

[1]The UV index I_{UV} is based on the spectral solar irradiance $E(\lambda)$ and is related to the erythemal effective irradiance E_{CIE}. The value is calculated using the constant factor $k_{er} = 40$ m^2 W^{-1} and the CIE erythemal action spectrum $s_{CIE}(\lambda)$ of equation 3.11 at page 112:

$$I_{UV} = k_{er} \cdot \int_{250\,\text{nm}}^{400\,\text{nm}} E(\lambda)\, s_{CIE}(\lambda)\, d\lambda$$

Figure 2.1: From left to right: warning sign for optical radiation, warning sign for electricity, mandatory action sign for eye protection, mandatory action sign for keep locked. At `http://en.wikipedia.org/wiki/User:DrTorstenHenning/imagegallery` these and other signs according to DIN standard 4844-2 are available both as vector graphics and bitmaps. These images are in the public domain.

match the wavelength emitted by each particular laser. Recommendations relating to warning signs, given above, apply equally to lasers and UV lamps.

2.1.3 Risks related to electrical power

Using mains power for either instruments or lamps, involves a risk of electric shock and this is especially true in humid conditions and outdoors. All components, like connectors, enclosures, and switches, should have an IP rating[2] and any other codes required for their use in a particular environment. When using mains power in a room where there is water, it is usually a requirement to have grounded mains sockets protected with residual current devices (RCD). These sense the current imbalance in the power line caused by a current leak to ground, for example caused by the accidental flow of current through the body of an operator. When this residual current is sensed, the device trips and disconnects power in a few milliseconds so preventing serious injury. RCDs and earth grounding should be regularly tested. Never use an electrical device that has a grounded plug in a mains socket lacking ground contacts because this will create a risk of electrocution and in addition may negatively affect the functioning of measuring instruments.

2.1.4 Safety regulations and recommendations

The EU directive 2006/25/EC about protecting workers from optical radiation can be found at `http://eur-lex.europa.eu/LexUriServ/LexUriServ.do?uri=OJ:L:2006:114:0038:0059:EN:PDF`. This directive does not apply to solar radiation. It gives maximum allowed levels of exposure from artificial optical radiation sources and other requirements on how to achieve these and

the monitoring of workers health. Britain's Health and Safety Executive (HSE) has produced a guide on how to apply this directive (not approved yet by EU), that can be found at `http://www.hse.gov.uk/radiation/nonionising/optical.htm`. At this address, there is also information on sun exposure indicating that 'UV radiation should be considered an occupational hazard for people who work outdoors'. The World Health Organization (WHO) has issued general recommendations about protection from the sun, which can be found at `http://www.who.int/uv/sun_protection/en/`.

2.2 Artificial sources of UV radiation

2.2.1 Lamps

2.2.1.1 Fluorescent lamps and tubes

Fluorescent lamps and tubes are low pressure mercury vapour lamps. The mercury vapour emits radiation at specific spectral lines, mostly in the UV region of the spectrum. Except for the case of germicidal UV-C lamps the inside of the glass tube is coated with a layer of 'phosphor'[3] which absorbs UV radiation and re-emits the energy as fluorescence, at longer wavelengths, either as UV-B, UV-A, or visible radiation. There are even some fluorescent tubes producing far-red radiation around 750 nm. For environmental UV studies we are interested in UV-B and UV-A emitting lamps. These lamps emit across a rather broad wavelength band, and have several minor secondary peaks in the UV-C and visible regions. Figure 2.2 shows the emission spectra of some UV lamps.

It should be remembered when designing experiments that UV-B lamps should be always filtered with cellulose diacetate film or some other material to remove UV-C, otherwise the effects observed will not be only due to

---

[2]IP stands for 'ingress protection' and usual codes are composed of the letters 'IP' and two digits. The first digit indicates level of protection for solids and the second digit level of protection for water. For example IP20 indicates protection from ingress of fingers, and no protection from water; IP54 means dust and splashing water protected; IP65 means dust tight and protected from water jets; IP67 adds protection to immersion in water to up to 1 m depth.

[3]Different phosphorous chemical compounds which fluoresce at different wavelengths, or their mixtures, are used in fluorescent lamps to produce radiation of different wavelengths. These phosphorous compounds are called 'phosphors' in the lamp industry.

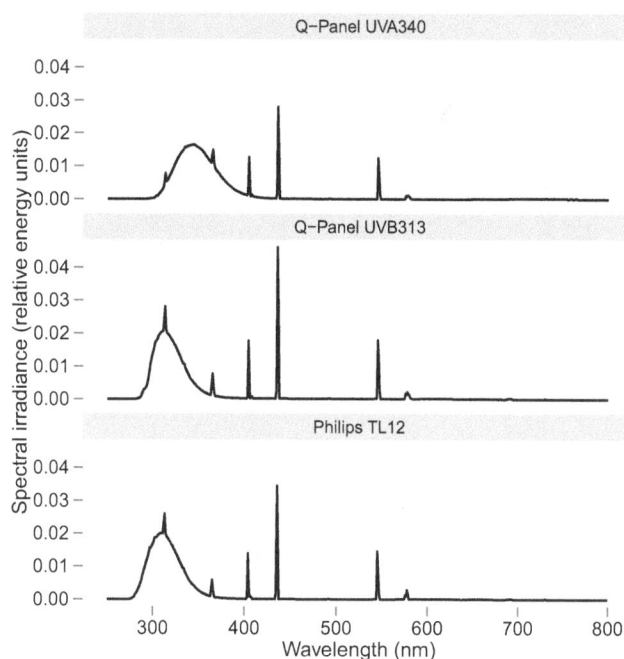

Figure 2.2: Emission spectra of unfiltered Q-Panel UVA-340, Q-Panel UVB-313 and Philips TL40/12 lamps. Spectral energy irradiance relative to total area under each curve. Measured with a Macam double monochromator scanning spectroradiometer.

UV-B but also to UV-C radiation. The emission of UV-C radiation might look small in energy terms but, being very effective in eliciting biological effects, it is capable of completely distorting the apparent response of plants and other organisms to UV-B radiation in an experiment. In the case of UV-A lamps, both UV-B and UV-C radiation should be removed by filtration.

Ultraviolet emitting fluorescent lamps also emit a small amount of visible radiation, even in the red and orange regions of the spectrum. This should not be forgotten when using these lamps as the only source of radiation. For example the very small amount of red-orange light can be enough to enhance germination of silver birch seeds in Petri dishes when irradiated with UV-B lamps (Pedro J. Aphalo, unpublished data).

The most commonly used UV-B and UV-A lamps are 1200 mm-long tubes, rated at 40 W of electrical power. These lamps are sold for materials testing (Q-Panel UVB-313 and UVB-340, Q-Labs Inc., Farnworth, England) and for medical use (Philips TL-40 W12, and TL-40 W 01, Philips, Amsterdam, The Netherlands). The second of these lamp types from Philips emits UV-B over a very narrow peak, and are not suited for simulating ozone depletion, but might be useful when the aim is not to simulate solar radiation. There are also other lamp types available, for example small-sized compact fluorescent

lamps such as PL-S 9W 01 from Philips.

Using fluorescent lamps requires some additional equipment (Figure 2.3). The main component is a "ballast" which limits the current through the lamp. Traditional ballasts are electromagnetic, comprising a coil wound on a ferrous core, and require one more component, a 'starter' for turning on the lamps. Lamps driven by such ballasts run at power-line frequency, 50 Hz in Europe and 60 Hz in USA, causing UV emission to flicker in time. This flicker is not visible to humans, but is visible to some insects. The modern alternatives are electronic ballasts driving the lamps at high frequency (in the order of 50–100 kHz) which together with the latency of the phosphor yields an almost constant radiation output from the lamps. Consequently, we recommend the use of high frequency ballasts to avoid artifacts. Electronic ballasts do not require starters.

Some types of electronic ballasts are dimming ballasts, allowing the adjustment of lamp output from full power down to a ballast-type-specific low value, usually somewhere between 1 and 10% of full power. Dimming of lamp output can also be achieved using electromagnetic ballasts and phase-angle dimming controllers but these enhance the visible flicker of the lamps. Some ballasts are designed for newer T8 (thin, 26 mm diameter) tubes rather than the old-fashioned T12 (thick,

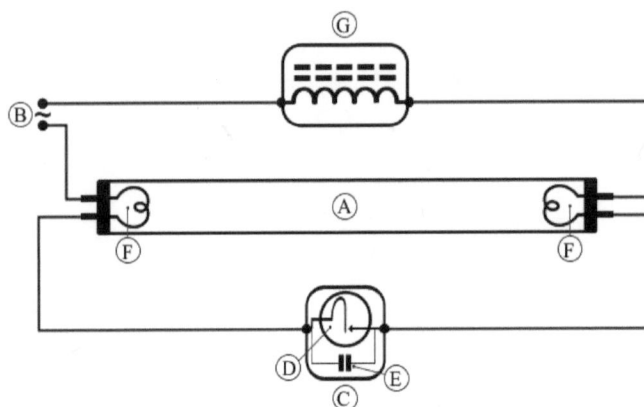

Figure 2.3: Diagram of a fluorescent tube and auxiliary equipment. A preheat fluorescent lamp circuit using an automatic starting switch. A: Fluorescent tube, B: Mains power (a.c.), C: Starter, D: Switch (bi-metallic thermostat), E: Capacitor, F: Filaments, G: Ballast. Image in the public domain, original file at `http://commons.wikimedia.org/wiki/File: Fluorescent_Light.svg`.

38 mm diameter) tubes used for most UV-B lamps, which may cause some problems with dimming, and affect the lowest power level achievable. Dimming ballasts are controlled digitally (for example DSI system for Tridonic PCA 1/38 T8 ECO, Tridonic GmbH & Co KG, Dornbirn, Austria) or by a direct current voltage signal (for example 1–10 V for Quicktronic HF 1X36/230-240 DIM UNV1, Osram, Munich, Germany). Dimming ballasts can be useful for adjusting doses for different treatments, or when building modulated UV-B supplementation systems (see section 2.2.6 on page 48).

Radiation output from fluorescent lamps varies markedly with ambient temperature as shown in Figure 2.4. The data in this figure were obtained by varying the temperature in a growth chamber where the lamps were located, and measuring the spectral irradiance with a spectroradiometer maintained at near constant temperature outside the chamber. Only the optical fibre entered the chamber through an instrumentation port. In non-modulated systems, and in experiments with treatments at different temperatures this should be taken into account and the irradiance from lamps should always be measured at the same temperature as during use of the lamps in the experiments. When the ambient temperature fluctuates, the lamp output should be measured continuously, or at least at a range of temperatures and irradiances estimated from a continuous record of temperature.

UV-B irradiance should never be adjusted by using different plant to lamp distances in different treatments. Changing the distance modifies the shading effect of lamp frames, and even small differences in PAR can create spurious differences between UV-B treatments (see Flint et al., 2009, for details). Furthermore, it is essential

to have controls both under frames with unenergised lamps and also under frames with energized lamps but filtered with polyester film to remove the UV-B, as the only difference between UV-B and non-UV-B treatments should be the irradiance of UV-B (see Newsham et al., 1996). If we want to assess the effect of shading by lamp frames alone an *additional* control with no lamp frames could be added to an experiment.

Fluorescent tubes are normally held on metal frames, and each frame supports several lamps. Because these tubes are long and narrow, using single tubes would provide a very uneven irradiance field. Care should be taken when choosing the spacing of the tubes, and the vertical distance between the tubes and plants. The model of Björn and Teramura (1993) can be used to simulate the spatial distribution of irradiance or fluence rate under such an array of lamps (for a reimplementation of the model as an R package see section 2.8 on page 67). Figure 2.5 shows the results of six such simulations, three for tubes evenly distributed along the frame ('equidistant') and three for tubes arranged following the projection on a plane of what would be equidistant distribution around the perimeter of a half circle ('cosine' pattern). The closer together the contour lines are, the steeper the change in irradiance. The examples in Figure 2.5 are for irradiance, and these results do not apply to fluence rate. For each size and number of lamps and size of frames, it should be possible to optimize either the evenness of irradiance or of fluence rate by means of the model (Figure 2.6). The unevenness of the UV radiation field should be taken into account when deciding where to the place plants, and also when measuring UV exposures under the frames.

When using UV lamps to supplement solar radiation,

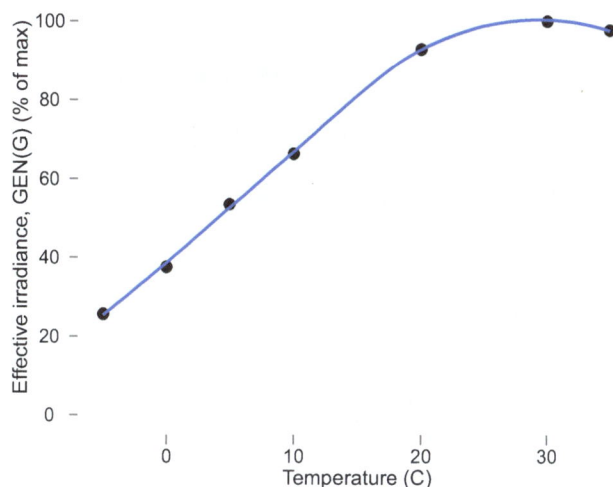

Figure 2.4: UV output of a lamp as a function of ambient temperature. Cellulose diacetate filtered Q-Panel UVB-313 lamps. GEN(G) used as BSWF. Doses calculated from spectral scans done with a Macam double monochromator spectroradiometer maintained at near-constant temperature. Redrawn from Aphalo et al. (1999).

shading by the lamps and their supporting structures can become a problem. This is discussed in section 3.10.3 on page 102 in relation to effective doses and enhancement errors.

2.2.1.2 Xenon arc lamps

Xenon arc lamps are specialised light sources that produce intense visible and UV radiation from an electric discharge in high pressure xenon gas. The lamps comprise tungsten electrodes inside a quartz envelope and produce an intense plasma ball at the cathode at 6000–6500 K temperature generating a spectral output throughout the visible and UV (Figure 2.7). Xenon arc lamps must be operated in a fully enclosed housing (Figure 2.8) that usually provides mirrors and focussing adjustments, electrical ignition and ventilation to cool the lamp. Xenon arcs provide intense point sources but with careful operation and adjustment are sometimes useful for experiments in photobiology. They are extremely strong sources and have a very high heat output in the infra-red waveband. This heat output typically requires their use with a water filter and/or a dichroic mirror reflector to reduce heating if used in experiments with plant material. A dichroic mirror transmits infra-red onto a heat sink and reflects only the required waveband (e.g. UV radiation) onto the target. After removal of excessive heat using a water filter or dichroic mirror they can be used with narrow band-pass filters for the determination of effects at different wavelengths such as action spectra. Care must be

taken not to focus the source onto the mirror or filter which can be readily damaged by excessive heat.

In addition to the essential safety considerations when using any UV sources (detailed in section 2.1), important additional safety precautions are necessary when using xenon arc lamps. The high pressure inside the lamps creates a risk of explosion during installation and during operation, which increases with lamp age. An impact resistant face shield and heavy protective gloves should be worn when handling a lamp and it must never be touched with the fingers as grease marks increase the risk of failure. Lamps are supplied in a protective cover which should remain in place until installation and they should only be used inside a specialised housing that provides explosion protection to contain flying glass should an explosion occur. A specialised lamp housing is normally provided with a fan for cooling the lamp which gets extremely hot and ventilation should continue for at least five minutes after the lamp is switched off. This is usually achieved in a specialised housing with temperature sensors or timers. However, some lamps (sometimes called 'UV-enhanced') produce short UV wavelengths (<250 nm) that generate large quantities of ozone which is toxic and can cause respiratory problems and asthma attacks. The ozone must be removed from the room and building by additional ventilation systems or absorbed and destroyed by special filters. Alternatively, 'ozone-free' lamps are most often used and have a quartz envelope that absorbs below 250 nm and so prevents excessive ozone formation. A xenon arc

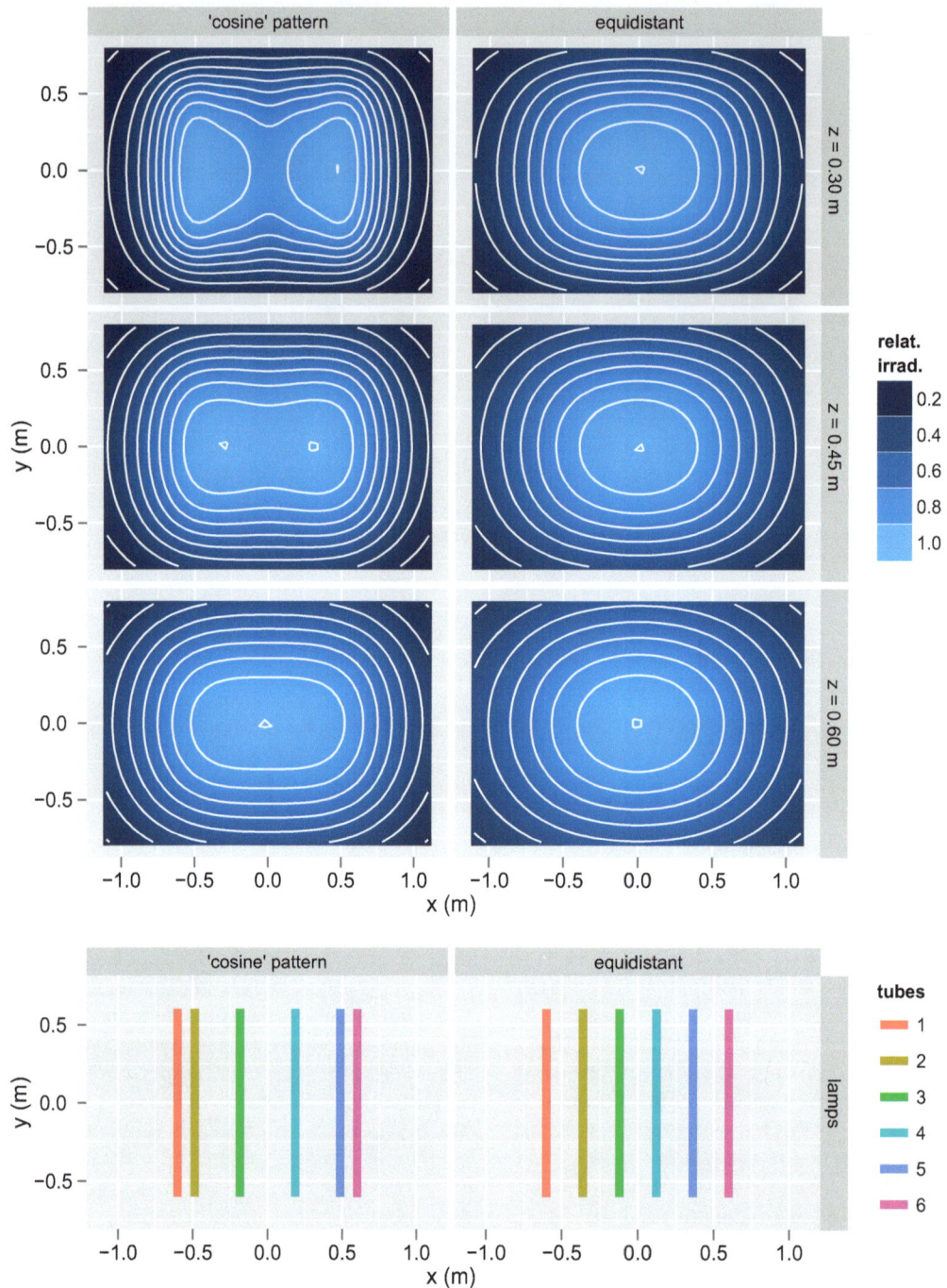

Figure 2.5: Irradiance field under lamps, arranged according to two different spatial patterns, 'cosine' and equidistant, at three different vertical distances below the lamps (z). For these examples the length of the frames (x) was set to 1.2 m and the length of the lamps (y) also to 1.2 m. The value used for y is that of the length of a 40 W or 36 W fluorescent tube. The calculations were done in R based on the algorithm of the model of Björn and Teramura (1993). The white contour lines are at a distance such that relative irradiance differs by 0.1 (10%) of the maximum irradiance. The data were normalized to the maximum irradiance for each separate panel in the figure. The two bottom panels show the positions of the lamps using the same scale as the upper six panels.

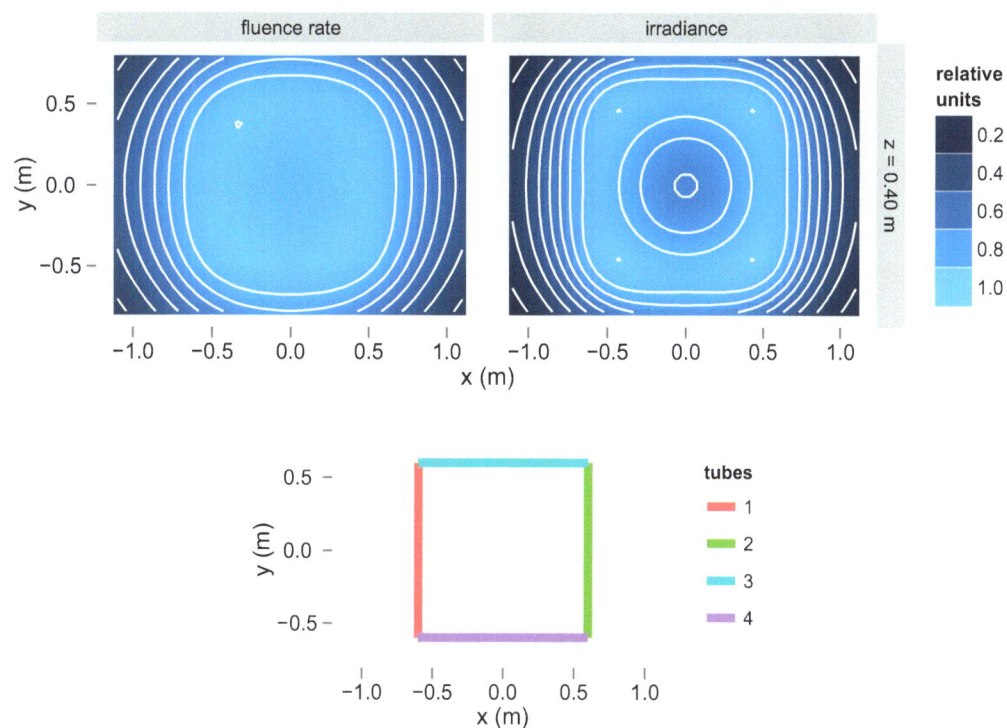

Figure 2.6: Irradiance and fluence rate fields under four lamps, arranged in a square, at a vertical distance below the lamps (z) of 0.40 m. For these examples the length of the frames (x) was set to 1.2 m and the length of the lamps (y) also to 1.2 m. The value used for y is that of the length of a 40 W or 36 W fluorescent tube. The calculations were done in R based on the algorithm of the model of Björn and Teramura (1993). The white contour lines are at a distance such that relative irradiance or fluence rate differs by 0.1 (10%) of the respective maximum. The bottom panel shows the positions of the lamps using the same scale as the upper two panels.

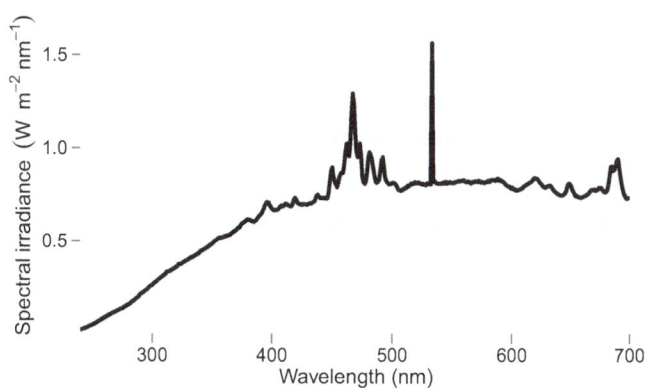

Figure 2.7: Spectral irradiance from a xenon arc source (Mueller Electonik-Optik, Germany) using an Osram 1000 W 'ozone-free' lamp after filtration through a water infra-red filter.

Figure 2.8: Xenon arc lamp (1000 watt, Mueller Elektronik-Optik, Germany) showing protective housing, high voltage power supply, water filter, beam turner, filter holder and water cooled cuvette with quartz glass window (School of GeoSciences, University of Edinburgh, UK). Photograph: Andy R. McLeod.

lamp is extremely bright and the intense visible radiation can damage the eyes in addition to UV effects. The arc or its intense image must never be viewed directly as it can cause permanent eye damage. When using the equipment, the illuminated areas may also be extremely bright and dark coloured goggles should be worn, such as those used in welding that also provide UV protection, ensuring that the eyes are also protected from side illumination as well as directly. It is essential to follow the manufacturers' instructions with respect to lamp life and operating current in order to maximise their useful life and minimise the risk of failure due to explosion. Further information on xenon arc sources and other high intensity lamp sources sometimes used in photobiology is provided by Holmes (1984).

2.2.2 Deuterium lamps

High intensity water-cooled deuterium lamps (150 W) have a fairly flat radiant intensity curve in the UV-B region (approximately 1.5–3.0 mW m^{-2} nm^{-1} between 280 and 300 nm at a distance of 30 cm) that is appropriate for mechanistic plant UV photobiology studies. This type of radiation sources (e.g. produced by Hamamatsu Photonics) requires a dedicated control box for operation but is still an affordable alternative to lasers, starting at about 2,500€ for a setup. High intensity deuterium lamps have been used for wavelength dependence measurements both in photobiology (Kalbin et al., 2005; Kalbina et al., 2008) and photochemistry (Kalbin et al., 2005) when fitted with the appropriate filters. The lamp has to be mounted horizontally to allow efficient water cooling but using spreading lenses and a mirror mounted in a 45° angle to the incident radiation a full *Arabidopsis* rosette

can easily be irradiated (Kalbina et al., 2008) and, the UV irradiation can be supplemented with PAR from external sources as desired. A drawback of this type of equipment is the short life of the radiation source (a few hundred hours). Spare lamps cost approximately 1 500 €.

2.2.3 LEDs

The principle of a light emitting diode (LED) is totally different to conventional incandescent light bulbs. It is a semiconductor consisting of two types of layer: one layer releasing electrons as a direct current to fill holes in another. Energy is released as a photon at a wavelength (colour) that depends on the band gap energy of the materials comprising each of the two layers (Figure 2.9).

Due to this emission of photons at a very precise wavelength, the typical spectrum of a LED shows one distinct peak. The emitted wavelength range can be increased by adding a fluorescing material to the LED. This is shown in Figure 2.10 together with a measurement of the solar irradiance at the Helmholtz Zentrum München in Neuherberg, Germany. The maximum spectral irradiance of the LED peak (measurement at 20 cm distance) is more than one order of magnitude lower than that of the solar irradiance (which is the reason a semilogarithmic scale is used). All measurements were made using a double-monochromator spectroradiometer (Bentham, Reading, UK).

If the correct semiconductor material is chosen, almost any wavelength can be produced even in the UV range. The intensity of UV LEDs (Figure 2.11) is much lower than that of LEDs emitting visible light, even at a shorter distance than in Figure 2.10. Normally, the optical power of a LED is around 1 W or less for visible radiation and

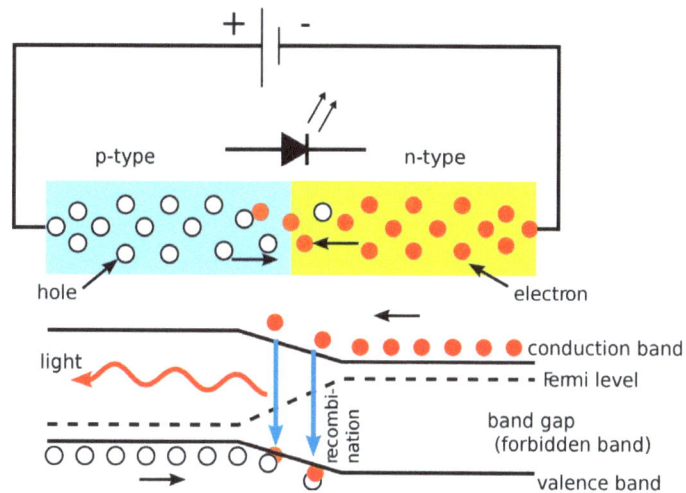

Figure 2.9: Light emitting diode (LED) junction and the flow of electrons and holes. Diagram by s-Kei, under Creative Commons Attribution-Share Alike 2.5 Generic license, original at `http://commons.wikimedia.org/wiki/File:PnJunction-LED-E.svg`.

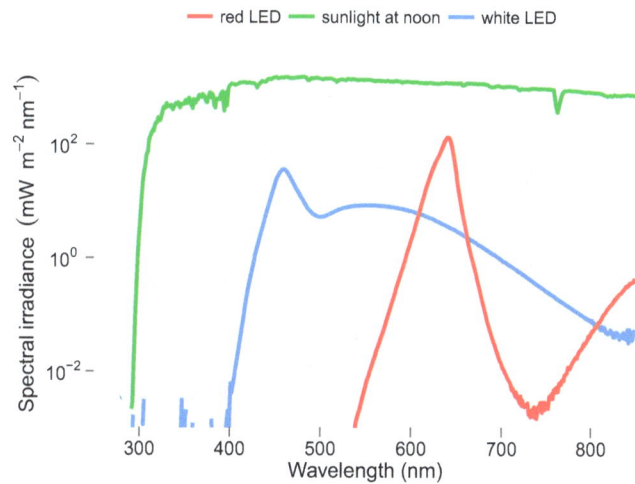

Figure 2.10: Emission spectra of a red and white LED measured at a 20 cm distance compared to an outdoor measurement at the Helmholtz Zentrum München, Neuherberg, Germany.

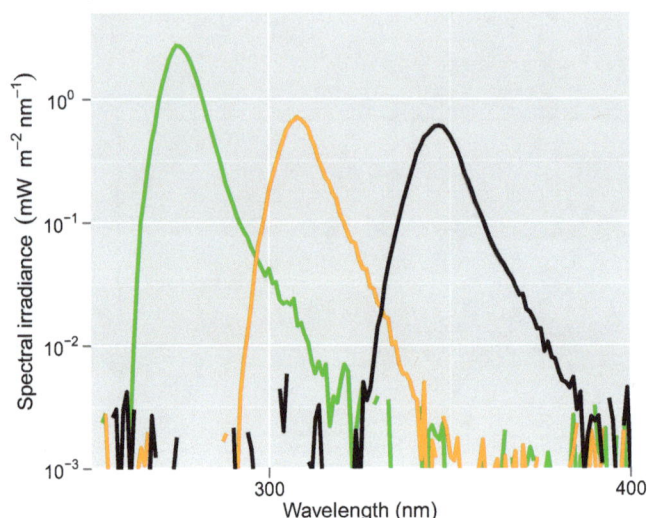

Figure 2.11: Spectra of three different LEDs emitting in the UV range (IMM Photonics GmbH, Unterschleißheim, Germany). The peak emissions are at 274, 308, and 347 nm and measurements were done at a 10 cm distance. The y-axis has a logarithmic scale, and the noise close to the x-axis is measurement-error noise, and not true emission by the LEDs.

only around 0.5 mW for UV radiation. There are LEDs available with higher power (many of them are arrays of tens of chips on a single package) but they require an additional cooling device. These arrays can emit up to 50 mW of UV radiation, however, they are very expensive. This is one of the most important reasons why LEDs alone are not useful *at the moment* for exposing plants to UV radiation. In addition, many different LEDs or specially designed LEDs are necessary to approximate the wavelength distribution of the solar spectrum. On the other hand, LEDs have been in common use for photosynthesis research for a long time (Tennessen et al., 1994) and many gas-exchange instruments use LEDs as a light source.

LEDs are low voltage direct current (DC) devices. The current through a LED must be limited. The simplest current limiter is a resistor connected in series with the LED. The calculator at `http://ledcalculator.net/` can be used to determine the value of the required resistor. There is also a handy pair of Android applications, 'elektor LED Resistor Calculator' and 'elektor Resistor Color Code', available for free from Google Play (`https://play.google.com/`).

2.2.4 Spectrographs

A spectrograph composed of a light source and a monochromator may be used in applications requiring spectrally-resolved UV radiation exposure of biological specimens. This may be the case when searching for

wavelength dependencies in plant responses or investigating specific effects known *a priori* to depend on wavelength.

The selection of a light source is application driven and depends on the requirements imposed by the study. The main requirements concern the intensity and the spectral distribution of the radiant output of the lamp. The geometry of the setup, including the source-target-distance and the area of exposure, sets certain limits not only on the light source but also on the characteristics of the monochromator.

A schematic representation of a typical single monochromator is shown in Figure 2.12. The light emitted by the light source is directed onto the entrance slit of the monochromator. The collimating mirror reflects the light onto the grating that diffracts the light into its spectral components. The diffracted light is reflected by the focusing mirror onto the exit slit. Two single monochromators may be aligned in a way that the exit slit of the first monochromator serves as the entrance slit of the second monochromator. This arrangement makes up a double monochromator which provides better stray light rejection than single monochromators. Commercially available monochromators of both types exist.

In the setup of a single monochromator shown in Figure 2.12, photons of only one wavelength would ideally exit the exit slit. However, as the slits have finite dimensions, the actual output would be a narrow line of radiation, peaking around a certain nominal central wavelength. Two approaches are commonly adopted to achieve the

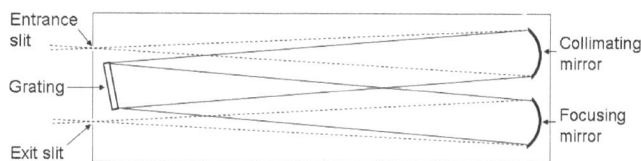

Figure 2.12: Optical setup of a single monochromator spectrograph.

production of multiple spatially-separated lines. The grating may be equipped with a drive arm rotating the grating. Ideally, for each position of the grating, only radiation of a certain wavelength is reflected and focused onto the exit slit one at a time. Alternatively, the focusing mirror may be omitted and replaced by a sample exposure plane. Ideally, in each position on the plane, only radiation of a certain wavelength would be present (Figure 2.13).

The main characteristics of a monochromator that should be considered include wavelength region, wavelength accuracy, bandwidth of the lines, and the amount of stray light. These characteristics set guidelines for design of the layout of the optical components, selection of the slit widths, and formulation of the specifications for the grating. An unavoidable compromise must be made between the wavelength resolution and the intensity. Finer separation of the wavelengths comes with lower intensities and vice versa. Stray light rejection may be improved by placing suitably-designed baffles inside the enclosure of the spectrograph. For applications requiring superior stray light rejection, a double monochromator should be used.

The main characteristics of the grating include groove spacing and a ruled area. For a spectrograph operating at UV radiation wavelengths, a holographic grating is a more feasible choice than a ruled grating because it can achieve a higher resolution. The main difference between ruled and holographic gratings is the process of their manufacture. A ruled grating has grooves scribed by a diamond on a ruling machine, whereas holographic gratings are produced by a photolithographic process using lasers which are able to achieve a higher groove density. Should the focal plane be flat, a concave grating must be used. The size of the grating has certain practical limits set by the manufacturers. The maximum size of the exposure area is essentially limited by the size of the grating.

Selection of the light source is defined by the desired wavelengths, the intensity levels, and the spectral distribution of the radiation. For experiments imitating natural exposure, light sources exhibiting radiative characteristics resembling those of natural sunlight, such as xenon-arc lamps, should be used. The heat tolerances of the

substrates of the gratings are limited. As a consequence, the light entering the monochromator may have to be filtered to remove excessive infra-red radiation which is often achieved using a water filter. (see subsubsection 2.2.1.2 on page 39)

In addition to the small spectrographs described above, there are also much larger instruments. Many of these dispense with the exit slit, and project the dispersed spectrum onto a 'stage'. Watanabe et al. (1982) describe a large spectrograph designed for irradiating biological samples. Over a curved 10 m-long focal plane covering wavelengths of 250 to 1000 nm, many samples can be irradiated simultaneously. There are few such spectrographs in the world, but they are important for measuring action and response spectra (e.g. Saitou et al., 1993).

2.2.5 Lasers

Laser stands for 'Light Amplification by Stimulated Emission of Radiation'. Stimulated emission can be induced in many different materials, and when light is confined in a cavity between mirrors the stimulated emission is amplified yielding a narrow beam of spatially- and temporally coherent, and collimated light. Lasers usually produce very narrow and intense beams of monochromatic light, although there are exceptions. Some lasers are tunable, meaning that the wavelength of the emitted light beam can be varied. One of the many ways of achieving this is to replace the mirror at one end of the cavity with a movable prism or grating. The lasing medium can be a solid crystal, a gas or a dye in solution. Lasers are said to be 'pumped' by a source of light, for example a lamp external to the lasing cavity.

Laser diodes are semiconductor lasers. They are solid state devices producing a laser beam based on the same principle as other lasers. They are pumped by the light emitted when electrons and 'holes' interact in the semiconductor junction. They are used to excite other types of lasers, in printers and CD/DVD/BD players. Blu-ray disc players use laser diodes which emit blue-violet light. There are UV emitting laser diodes of low power, for example, emitting 20 mW at 375 nm. The semiconductor material used in this case is GaN. In addition the cata-

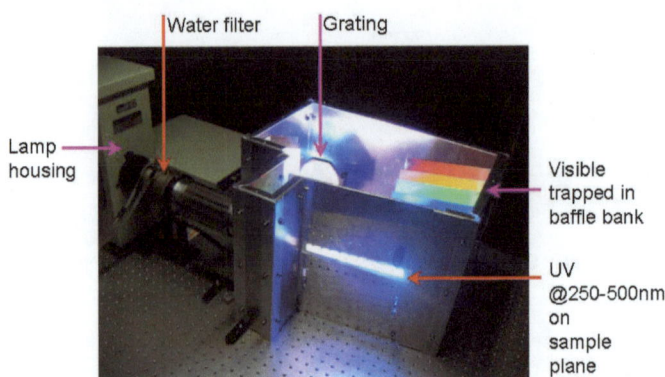

Figure 2.13: Spectrograph constructed to produce spectrally-resolved UV radiation for material degradation studies. (Cover open to show components.) Photograph: Anu Heikkilä.

logue of Roithner Lasertechnik GmbH (Vienna, Austria) lists 'diode pumped solid state lasers' emitting at 266 nm and at 355 nm. Non solid-state lasers that emit in the UV region are also available. For example argon ion lasers continuously emit at wavelengths of 334 and 351 nm.

The power of lasers varies from 1–5 mW for laser pointers to 100 kW for lasers under development for military use. The coherent and concentrated beam can cause serious damage to eyesight, and safety precautions should be taken during their use, unless they are of very low power (see section 2.1 on page 35).

Lasers can be used for example in characterizing the slit function of spectrometers (see section 3.7.2.2 on page 93), for aligning optics, and in the ubiquitous barcode readers used in supermarkets and laser pointers used by speakers during lectures. They are used in instruments like confocal microscopes and different analytical procedures in chemistry and biology. They can also be used to excite photochemical or photobiological systems if the target is small.

For the purpose of UV photobiology, tunable optical parametric oscillator (OPO) pump lasers (pump wavelength 355 nm) are especially useful in mechanistic studies such as in accurate wavelength dependence determinations or action spectroscopy (O'Hara, Strid and Jenkins, in preparation). Placing a cuvette holder in the beam, a plant extract or a protein solution can be accurately irradiated using 50 or 100 µl cuvettes. By using the appropriate optics (lenses, mirrors, etc.; Figure 2.14), the geometry of the beam can be manipulated so that a larger area or sample (several cm^2) can be illuminated. This is large enough for exposing for instance part of an *Arabidopsis* rosette, or for simultaneously irradiating two detached *Arabidopsis* leaves. Tuneable lasers can also be used for studies of complex mechanistic interactions

between several plant photoreceptors, since they can be used within a short time interval (seconds) at different wavelengths, e.g. for irradiation first in the UV-B, then in the blue or red.

When using lasers for photobiological purposes, a number of circumstances need to be taken into account. Lasers emit their radiation in the form of pulses and this should be kept in mind when designing the experiments. Usually, both the energy of the pulses and the pulse rate can be varied within limits, depending on the sophistication (and price!) of the equipment. Pulse rates between 1 to 20 Hz in the more affordable machines up to 1000 Hz in the more expensive ones can be obtained. The pulses are typically 5–10 ns in length and the linewidth is usually around 0.5 nm (Figure 2.15), making this type of instrumentation ideal for studies of the initial UV signalling events on the biochemical and cell biological levels. Also, due to the physical principles on which a tunable laser in the UV region relies, some residual radiation with double the desired wavelength may still be present in the beam exiting the apparatus. This has to be taken care of by using a filter in the light-path blocking the longer wavelengths. For instance, when irradiating a plant with a wavelength of 300 nm, care must be taken so that any 600 nm component will be blocked out to be sure that the biological effect is due to the UV-B alone and not the red light. Of course, the laser UV-B radiation can be supplemented with PAR from other light sources as required.

Laser radiation is measured as the energy emitted (in J) in each pulse or in a train of pulses using a pulse energy meter. This then has to be related to the area actually irradiated. The maximal output energy for a flash at 280 nm typically ranges between 40 and 200 µJ. The spectral bandwidth and the accuracy of the wavelength

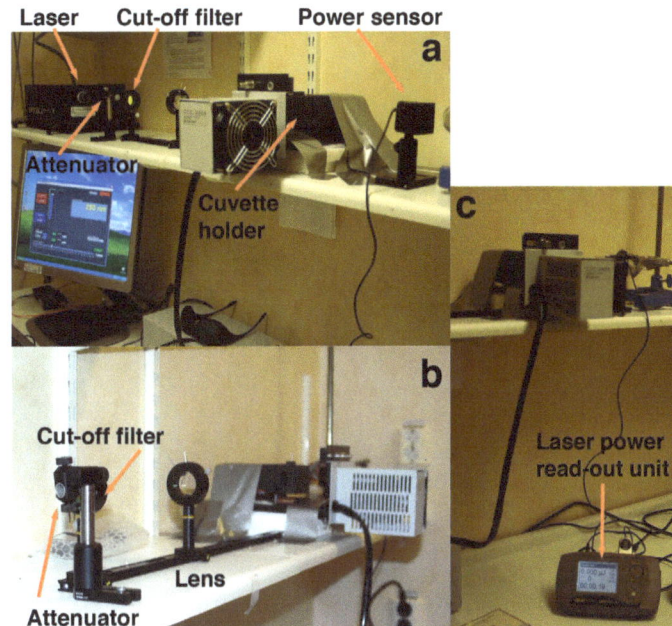

Figure 2.14: Set-up with an Opolette 355 II+UV tunable laser (a) for irradiation of plant extracts or a protein solution placed in a 50- or 100-μl cuvette in the Peltier-cooled cuvette holder (a–c). The Figures also show the attenuator, the cut-off filter (for removing red light), the lens for achieving the desired geometry of the beam, and the laser pulse power sensor (a, b), that all are present in the light path, as well as the read-out unit for the power sensor (c). For exposure of plant leaves or a small pot containing an *Arabidopsis* plant, the cuvette holder is removed and a mirror is placed at 45° angle where the power sensor is seen in (a). This mirror deflects the laser beam through a hole in the shelf hitting an area on the bench top where the leaves/plant are placed. Photographs: Åke Strid.

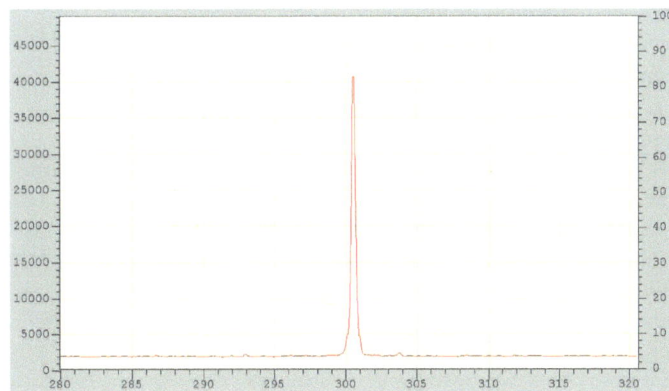

Figure 2.15: The sharp 300 nm peak from an Opolette 355II+UV (Opotek Inc, Carlsbad, CA) tunable laser with a halfbandwidth of 0.4 nm as measured using a SM440 diode array spectroradiometer (Spectral Products, Putnam, CT). Representation on the Y-axis is in arbitrary units. Figure by Strid, Åke (2012) unpublished.

setting should be checked prior to use with a calibrated diode array spectroradiometer (Figure 2.15). As might be expected, the major drawback of using lasers, in addition to the small area that usually can be irradiated, is the cost. Lasers useful for UV-B photobiological purposes at present start at approximately 50 000 €.

2.2.6 Modulated UV-B supplementation systems

There are two types of UV-B supplementation systems for use outdoors: square-wave and modulated. Square-wave UV-B supplementation systems work by simply switching the lamps on and off. The only control on the dose of UV-B is the length of time that the lamps are energized each day. Modulated systems work by continuously adjusting lamp UV-B-radiation output, which is usually called dimming. Furthermore, in modulated systems, the dimming of the lamps is adjusted by means of a feedback control system. UV-B radiation is measured both with UV-B supplement and without and the control system is programmed so that the treatment is a fixed percent increase in UV-B above the control condition, or in some cases, above true ambient conditions. McLeod (1997) reviews many of the early modulated and square-wave systems, and discusses many of the compromises and limitations involved in their use and design. Figure 2.16 shows one lamp frame from a modulated system, and Figure 2.17 shows an overhead view.

A modulated system automatically adjusts the UV-B supplement following changes in natural UV-B irradiance with time of day and also compensates for ageing of cellulose diacetate filters, and changes in lamp output with ambient temperature.

Caldwell et al. (1983) describe a modulated system based on custom built electronics, which works even at very low ambient temperatures. Aphalo et al. (1999) describe a system built from off-the-shelf components and controlled by a datalogger, which does not work at temperatures below 0°C. Figure 2.18 shows, for this system, the daily course of UV-B irradiance under near-ambient control and UV-B enhancement frames throughout two days with different cloud conditions. Occasionally, modulated systems have also been used inside greenhouses (e.g. Hunt and McNeil, 1998). Another approach is to use a proportional-integral-derivative algorithm in an industrial micro-controller module to control the dimming of the lamps. Such a modulated systems has been built based on Gantner intelligent modules (Gantner Instruments Test & Measurement GmbH, Darmstadt, Germany, Matti Savinainen, pers. comm.). It is also possible to use a personal computer and a PID control algorithm implemented in a graphical instrument control and meas-

urement language like LabVIEW (National Instruments Corporation, Austin, TX, USA) or FlowStone (DPS Robotics, UK).

Modulated systems, like any other UV-B supplementation system, require frequent checks and replacement of burnt lamps and aged filters. It is recommended to replace acetate filters regularly. For lamps (except for the odd 'early deaths') it is recommended that they are replaced following a fixed schedule as lamp output degrades with lamp age. This is especially important in those systems not having separate feedback control for each lamp frame, as having lamps of different ages in monitored and slave frames would lead to inaccuracies in the level of UV-B supplementation between frames.

Modulated systems are preferable to square-wave systems as they avoid excessive supplementation during periods of low PAR and low natural UV-B irradiance. They also compensate for fluctuations in lamp output. However, they are more difficult to design and build, and consequently more expensive. The drawbacks of square-wave systems can be moderated by completely switching off the system on rainy and cloudy days, or by automatically switching off UV-B supplementation when PAR irradiance goes below a certain threshold. The duration of square-wave exposure may also be limited to fixed hours either side of solar noon. Musil et al. (2002) evaluated the errors involved in the use of square-wave systems and concluded that they are not very large, whereas S. Díaz et al. (2006) concluded that modulated systems are preferable. However, it is difficult to assess the effect of the unrealistic treatments in square-wave systems on the outcome of experiments as comparisons have been solely based on differences in UV-B irradiation regimes rather than on the responses of plants.

Enhancement errors in the calculation of effective UV irradiance are discussed in section 3.10.3 on page 102.

2.2.7 Exposure chambers and sun simulators

Two main types of sun simulators exist: small systems based, for example, on xenon-arc lamps (see 2.2.1.2 on page 39), and large systems built by combining several types of lamps and filters to achieve a simulation of the solar spectrum; so called "sun simulators" and "(walk-in-size) exposure chambers". The latter are sometimes called a 'phytotron' (Bickford and Dunn, 1972). The principal concept of how to design plant growth chambers can be found for example in Langhans and Tibbitts (1997).

Here, a description is given in more detail of the phytotron built at the Research Unit Environmental Simulation of the Helmholtz Zentrum München, Neuherberg, Ger-

---

[4]Website: `http://www.helmholtz-muenchen.de/eus`

Figure 2.16: Modulated UV-B supplementation system in Joensuu in 1997. See Aphalo et al. (1999) for details. Photograph: Pedro J. Aphalo.

Figure 2.17: Modulated UV-B supplementation system in Joensuu in 2012. The new design of the support reduces shading. Infrared heating has been added along the middle of the frames. Photo: Matti Savinainen

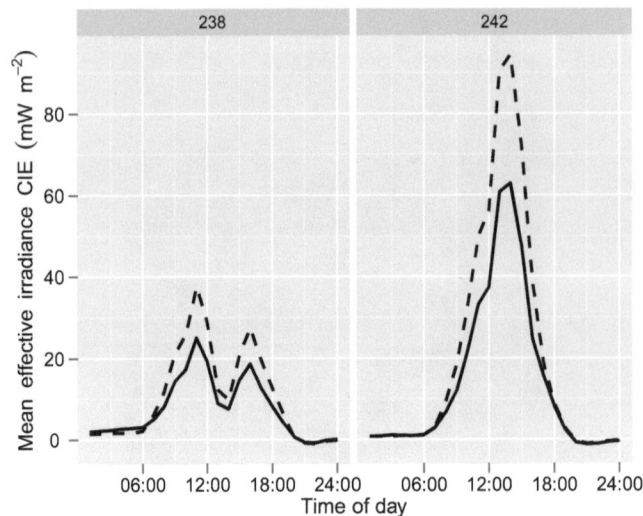

Figure 2.18: UV-B irradiance under a modulated system on two days differing in cloud cover. Data for 30% enhanced UV-B treatment (dashed line) and UV-A control (solid line). Days 238 and 242 of year 1997, at Joensuu, Finland. UV irradiance measured with two matched Vital Technologies' (Bolton, Ontario, Canada; no longer in business) '*Blue Wave*' BW-20 erythemal sensors. From Aphalo et al. (1999).

many[4]. This phytotron facility consists of a set of seven closed chambers (length · width · height):

- four walk-in-size chambers (3.4 m × 2.8 m × 2.5 m),

- two medium-size sun simulators (1.4 m × 1.4 m × 1.0 m),

- one small sun simulator (1.2 m × 1.2 m × 0.4 m).

In order to extrapolate response of plants from experiments to those in their natural habitat, these experiments need to be performed under realistic and reproducible conditions. The radiation provided must be as realistic as possible. Not only the quantity but also the spectral quality of radiation has to match the seasonal and diurnal variations occurring in nature (Caldwell and Flint, 1994a). This includes the steep absorption characteristics of UV radiation that result from the filtering of solar irradiance by stratospheric ozone as well as the balance between the UV-B, UV-A, and PAR. As no single artificial light source can simulate both spectral quality and spectral quantity of global irradiance, a combination of metal halide lamps (400 W HQI Daylight, Osram, Germany), quartz halogen lamps (500 W Halostar, Osram, Germany), and blue fluorescent tubes (40 W TLD 18, Philips, the Netherlands) are used in order to simulate the spectrum from the UV-A to the near IR wavelengths. Excess infra-red radiation is removed by a layer of water. Underneath this water filter, additional quartz halogen lamps (300 W Halostar, Osram, Germany) are installed to adjust the

mid and far IR radiation. The missing UV-B irradiance is supplemented by UV-B fluorescent tubes (40 W TL 12, Philips, the Netherlands). The radiation output of these fluorescent tubes, however, extends to well below 280 nm. This portion must be blocked very efficiently. Selected borosilicate and lime glass filters as well as acrylic 'glass' filters and plastic sheets provide a sufficiently steep cut-off at the desired wavelength. Different combinations of these glasses and films allow the cut-off wavelength to be altered, thus enabling a simulation various UV-B scenarios as shown for example in figure 1.20 in section 1.8 on page 26. The diurnal variations of the irradiance are achieved by switching appropriate groups of lamps on and off. The optimised lamp configuration of the ceiling is presented in Figure 2.19. A more detailed description is given by Seckmeyer and Payer (1993), Döhring et al. (1996) and Thiel et al. (1996). Figure 2.20 shows a picture of a sun simulator and the schematic outline.

The lamps are mounted far above the cultivation area in order to obtain a homogeneous spatial distribution of the radiation. Deviations from homogeneity, indicated by the ratio of spectral irradiance at any position to that at the central position, do not exceed 20% for all wavelengths but, due to non-symmetric lamp mounting, can depend on wavelength. The horizontal distribution of PAR in the cultivation area is shown in figure 2.21. The horizontal distribution of UV-B and UV-A radiation is similar and not shown here.

Figure 2.22 shows a comparison of the spectral ir-

■ (green)	metal halide lamp 400 W
■ (blue)	metal halide lamp 400 W with blue filter
■ (yellow)	quartz halogen lamp 500 W
■ (dark blue)	blue fluorescent lamp 38 W
■ (orange)	quartz halogen lamp 300 W (below water filter)
■ (grey)	auxillary lamp
	96 UV-B fluorescent lamps (not shown)

total electrical input 35 kW

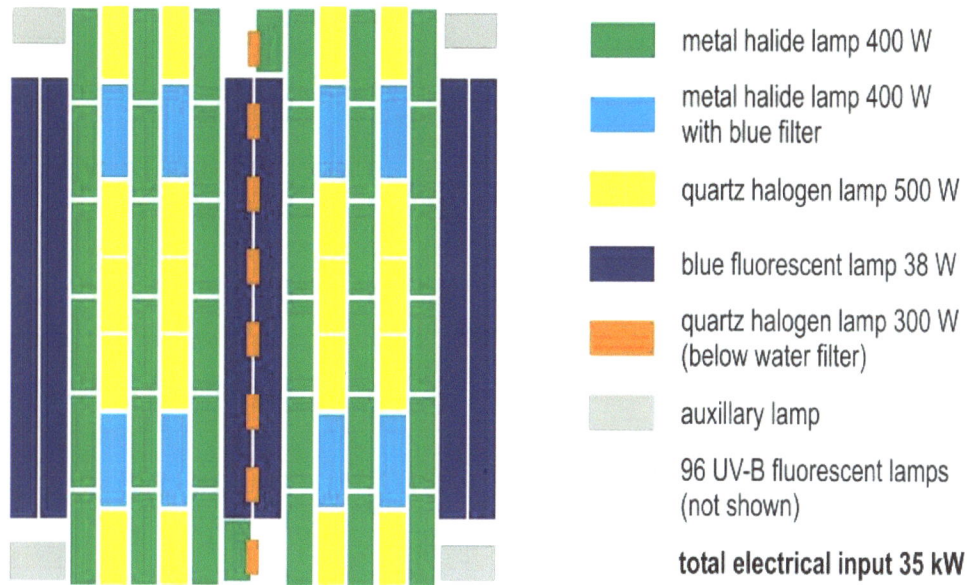

Figure 2.19: Optimised lamp configuration of the ceiling of a sun simulator at the Research Unit Environmental Simulation of the Helmholtz Zentrum München, Neuherberg, Germany.

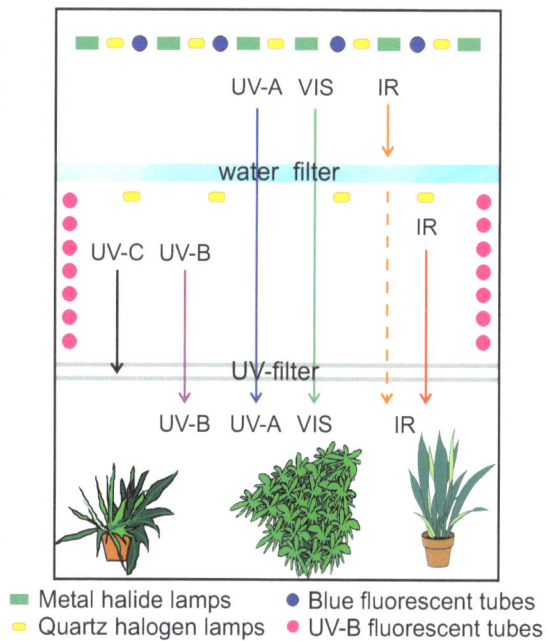

Figure 2.20: Picture and schematic outline of the lamp and filter configuration of a sun simulator at the Research Unit Environmental Simulation of the Helmholtz Zentrum München, Neuherberg, Germany. Photograph: Andreas Albert.

51

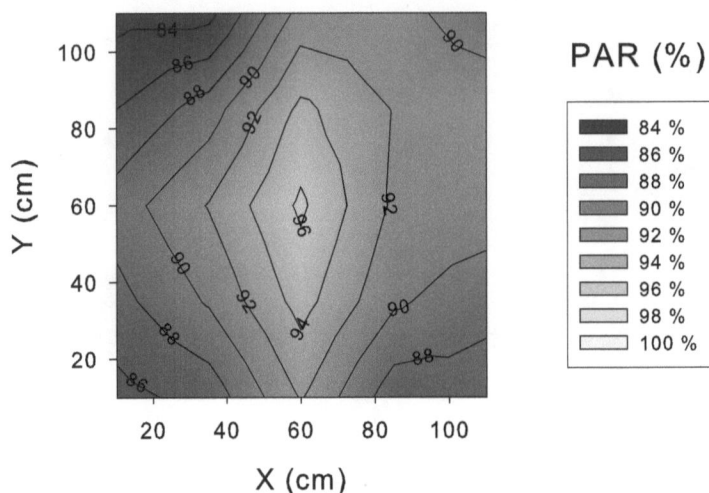

Figure 2.21: Horizontal distribution of PAR in the cultivation area of a sun simulator at the Research Unit Environmental Simulation of the Helmholtz Zentrum München, Neuherberg, Germany.

radiance of the small sun simulator, provided with a double layer of Tempax glass (Schott, Germany), and a typical outdoor spectrum measured at the field station of the Helmholtz Zentrum München, Neuherberg, Germany. Measurements in the sun simulator show that the steep, realistic shape of the UV-B edge and the UV-B:UV-A:PAR ratio can be simulated very close to nature. The UV-B:UV-A:PAR ratio of the sun simulator is 1 : 23 : 194 and matches the natural conditions of 1 : 25 : 206 very well.

Typical irradiance data from the phytotron compared to a field measurement (on 17 April 1996 at the Helmholtz Zentrum München, Neuherberg, Germany, solar zenith angle $\theta_s = 38°$) are listed in table 2.1.

Besides the lighting, temperature, and humidity, the atmosphere in the chamber is also controlled. Typical gaseous pollutants such as ozone, nitric oxides, and combustion residuals can be introduced. The effects of carbon dioxide and hydrocarbons on plants can also be studied. Modern control technology with central monitoring ensures a safe and well-defined operation.

2.3 Filters

In UV research, optical filters are used in different contexts, in measuring instruments and in UV radiation sources. In the case of instruments, they are used in most broadband sensors to achieve the desired spectral response. They are also used in some spectrometers to improve stray light performance and remove second order artifacts.

They are used in lamp sources like laboratory solar simulators based on xenon arc lamps. In experiments with UV-B fluorescent lamps they are used to absorb short wavelength radiation in the UV-B treatment and to achieve comparable no UV-B controls. These uses are discussed in section 2.2.1 on page 36. In the current section, in addition to introducing the optical properties of filters, we will mainly discuss the use of filters to manipulate the spectrum of sunlight in field experiments.

2.3.1 Optical properties of filters

There are different types of filters. Band-pass filters transmit radiation in a given range of wavelengths or band, and absorb or reflect light outside this range. The band can be either wide or narrow. Cut-off filters can be either long-pass, or short-pass: respectively transmitting all radiation of longer wavelengths than a cutoff wavelength and transmitting all radiation of wavelengths shorter than a cutoff (less common). The transition is a slope that can be steep or gradual depending on the filter.

Just like any object, filters reflect, absorb or transmit radiation. These are the only three possible fates of incident radiation. Reflectance (ρ), absorptance (α) and transmittance (τ) are the corresponding fractions of the incident radiation such that $\rho + \alpha + \tau = 1$ (or if expressed as percentages they add up to 100%). Absorptance should not the confused with absorbance (A) which is a different quantity given by $A = \log \frac{1}{\tau}$. All these quantities can be defined either for a broad band, or for very narrow band (e.g. $\tau(\lambda)$ where λ is the wavelength). In the latter case, we can measure across wavelengths obtaining a

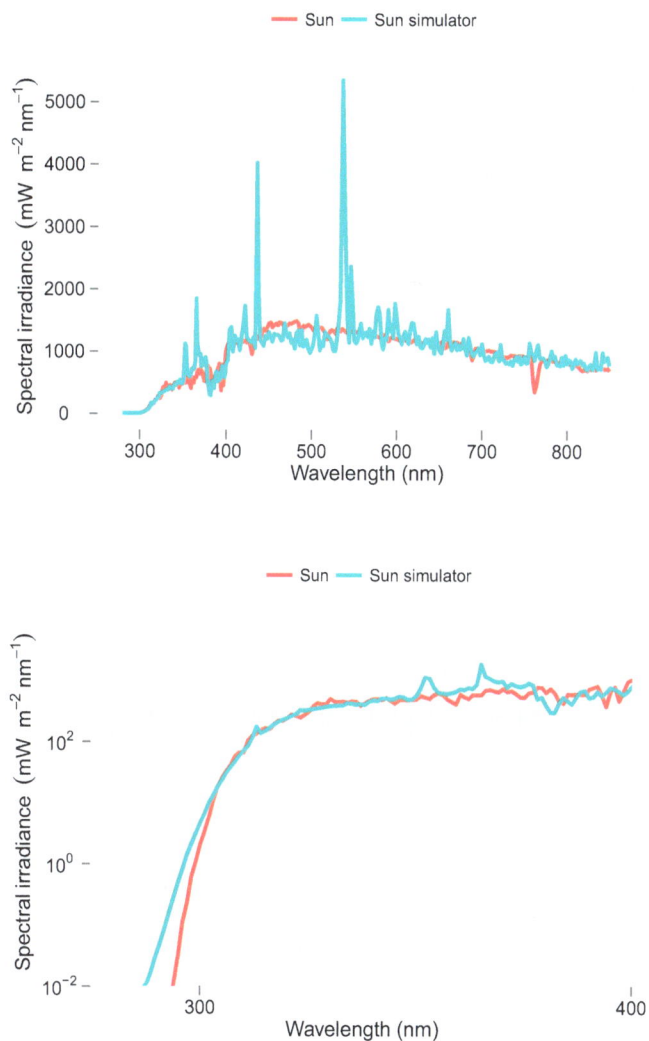

Figure 2.22: Spectral irradiance of the small sun simulator (blue) with a double layer of Tempax glass (Schott, Germany) compared to an outdoor spectrum measured (red) on 17/04/1996 at the field station of the Helmholtz Zentrum München, Neuherberg, Germany.

Table 2.1: Typical irradiance data from the small sun simulator (Small) and the walk-in-size chambers (Large) compared to a typical outdoor measurement at the Helmholtz Zentrum München, Neuherberg, Germany. The irradiance values of the wavelength ranges are given in units of W m^{-2}.

	Small	Large	Outdoor
E_{UVB}	1.53	1.20	1.10
E_{UVA}	54.9	22.7	47.2
E_{PAR}	430	250	390

spectrum.

When describing filters, the main property of interest is spectral transmittance. Most filters work by absorbing radiation of the unwanted wavelengths. However, some special filters work by reflecting the unwanted radiation. This is important when working with high power light sources: reflecting filters warm up much less that absorbing filters. An example of reflecting filters are the so called heat mirrors, which reflect infra-red radiation.

For light absorbing filters, like many of those made from plastic or optical glass, the optical characteristics depend not only on the material but also on its thickness. For this reason it is important to always indicate both type and thickness when describing a filter. In many cases one can take advantage of this phenomenon when planning experiments as the cutoff wavelength depends on the thickness of the filter.

Another important property of filters is whether they scatter radiation or not. When light is scattered the diffuse radiation component increases. For example, normal glass does not scatter visible radiation while opaline (white glass) does. As the proportion of diffuse radiation in PAR affects canopy photosynthesis (Markvart et al., 2010; Okerblom et al., 1992; Urban et al., 2007, 2012) one should use filters with similar scattering properties for all treatments.

Filters can be made from optical glass, gelatine, and many different synthetic organic compounds. In many cases, coloured substances are used as additives to the whole thickness of the material but sometimes a layer is merely applied to the surface or between two layers of transparent material. Glass filters tend to be very stable, but many plastics and the additives they contain react when illuminated, and deteriorate gradually when exposed to visible- and especially UV radiation. Filters made from optical glass are a more expensive than those made from plastics and tend to be readily available only in small sizes. Plastic films used as filters tend to be available in large sheets or rolls. Sometimes, liquid filters, usually aqueous solutions of chemicals, can also be useful (see Chapter 11 in Montalti et al., 2006, for spectral transmittance for several liquid filters). Sampath-Wiley

and Jahnke (2011) describe a new type of liquid filter which makes it possible to obtain a realistic simulation of the solar UV spectrum under laboratory conditions using normal UV-B fluorescent lamps.

The engineering quality of atmospheric UV absorption by glass filter techniques is sufficient for most plant experiments. However, this method has its limitations. Strong UV exposure causes rapid ageing of borosilicate glass filters due to a physico-chemical effect known as solarisation. This originates in UV-induced changes in the oxidation state of iron contaminants present in the glass matrix. The oxidation of Fe^{2+} to Fe^{3+} is accompanied by an absorption shift to longer wavelengths within the UV range. A decrease in the transmittance of fresh glass filters occurs within their first few hours of use, followed by a more-gradual long-term decline. Hence, changes in transmission during an experiment can be largely avoided by the pre-ageing of new filters. In addition, there is a gradient of ageing with the depth of the glass. Soda-lime glass exhibits very much reduced ageing compared to borosilicate glass. The solarisation of the soda-lime glass ceases after a few hours of UV treatment and this may be due to its lower content of iron contaminants (Döhring et al., 1996).

2.3.2 Manipulating UV radiation in sunlight

When using plastic films in the field they need to be supported in a way such that they remain in place even under windy conditions and also so that rainwater does not accumulate on top of them. It is important to be aware that plastic filters modify the microclimate in several ways. If they absorb infra-red radiation, even if transparent at other wavelengths, they will alter the energy balance of plants and soil under them. This produces a greenhouse effect that increases temperatures. If they absorb PAR they will affect photosynthesis, and even small differences in PAR transmittance (τ_{PAR}) between treatments could be important. Plants are partially protected from wind and completely shielded from rain by the filters. In some cases even plasticizer additives (different phthalates) added during plastic manufacture have been implicated in artifacts caused by the use of cellulose di-acetate film in UV-B exclusion experiments, particularly when ventilation is restricted (Krizek and Mirecki, 2004).

As filters have so many side effects, it is not surprising that a comparison against a no-filter control usually yields large differences, demonstrating that comparisons examining the effects of UV attenuation should always be done against controls under UV transparent films. Of course, we may want to know how similar the conditions under control filters are to natural conditions. In this case at least two types of controls are needed: a control with a UV transparent filter and a control without any filter. Figures 2.23, 2.24, 2.25 and 2.26 show some typical setups for potted plants and branches respectively. Filters on tree branches have been used by Rousseaux et al. (2004), Kotilainen et al. (2008) and others. Filters have also been used to cover patches of natural vegetation by Phoenix et al. (2003), Robson et al. (2004) and others. Filters have been used in experiments with potted seedlings or plants by Hunt (1997), Kotilainen et al. (2009), Morales et al. (2010), and others.

The spectral transmittance of several commonly used films is shown in Figure 2.27, and the resulting filtered sunlight spectral irradiance in Figure 2.28. For near-ambient-UV controls cellulose diacetate, polythene or polytetrafluoroethylene (PTFE) are normally used. Cellulose diacetate is not a good option as its optical properties deteriorate fast, and after deterioration it absorbs more UV-B radiation (Figure 2.29), tears easily, and is affected by water. Cellulose diacetate is used for removing UV-C from the radiation emitted by UV-B lamps (see section 2.2.1.1 on page 36). In contrast as there is no UV-C in solar radiation at ground level, other materials that are more durable can be used for near-ambient controls in attenuation experiments. Polythene film (types without UV absorbing additives only) is very stable, and cheap, but usually scatters light slightly more than the films used for UV-B attenuation. PTFE and related polymers (brand names Teflon, Hostaflon, etc.) are extremely stable, transmit UV radiation very well, and some types produce little scattering[5]. Also films made from polychlorotrifluoro-ethylene (PCTFE) (sold under the trade name ACLAR) can be used. For UV-B attenuation experiments, polyester film (e.g. brand names Mylar, Melinex, Autostat) is the filter material most frequently used for attenuating UV-B with only a small effect on UV-A irradiance. Sometimes soda glass (normal window glass) or special acrylic plates are used instead of polyester. Polyester film produces little scattering and glass and acrylic panes almost none. For attenuation of the whole UV band, UV-absorbing theatrical gels can be used. They are called gels for historical

reasons but nowadays either polyester or polycarbonate is used as base material. The most useful type is the one with code #226, available with similar specifications from several manufacturers (Rosco, Lee filters, Formatt filters, see the Appendix on page 68 for addresses). Rosco #0 is an UV-A and UV-B transparent theatrical gel with spectral transmittance rather similar to thin cellulose diacetate. Some polyester films can also remove all UV wavelengths (e.g. brand name 'Courtgard'). Some experimental greenhouse cladding films have cut-off wavelengths in the middle of the UV-A band and provide additional possibilities. Theatrical gels are available only in one standard thickness, but cellulose diacetate, polyester, PTFE and polythene films are available in several different thicknesses. Thickness affects both mechanical and optical properties (Figure 2.30). The most common thicknesses used for cellulose diacetate and polyester are 100 to 125 μm. For polythene, thinner films can be used to minimize scattering if mechanical strength is not limiting (e.g. 50 μm). As PTFE is a strong material, thin films can also be used, but one should be careful to match PAR transmittance between all treatments.

Filters can be mounted on wooden, metal or wire frames, or sometimes, especially when covering branches, on chicken wire net. It is important to carefully chose the fastening method. On wooden frames staples or thumb pins can be used, but the films tend to tear where they have been perforated so reinforcing the edges with transparent or duct tape may be necessary. It is important to use tapes that produce no toxic fumes, and to use the same amount and type of tape for all treatments (e.g. not use more tape for the films that break more easily). In harsh environments, the propensity for filters to tear or fracture can be greatly reduced by stretching them taut over the filtered plots to keep them still. This can be achieved by unrolling filters firmly attached to two cylindrical rods/poles clamped under tension to a structure anchored in the ground (Figure 2.24 on page 56), . This approach also has the benefit over wooden frames of enabling a large number of filters to be carried to the field at once undamaged, so can be of use to researchers working at inaccessible field sites. When mounting the filters good ventilation should be maintained, to avoid warming up of the plants and soil under them. Careful consideration is needed when making the compromise between the UV reduction achieved by adding plastic curtains to increase the filtration of scattered or 'diffuse' radiation (or direct radiation when the solar elevation angle is small at high latitudes), and the exacerbation of warming and further reduction in ventilation that this will cause. In long-term filtration experiments, the growth of dense vegetation

---

[5]Some types of PTFE scatter light almost perfectly and are used to manufacture cosine diffusers for broad band sensors and spectroradiometers. Those plastic films useful as filters in UV-B experiments are the non-scattering ones.

Figure 2.23: Filters in the field, mounted on wooden frames, and slightly tilted to avoid accumulation of rainwater. Viikki campus, University of Helsinki, Finland. Photograph: Pedro J. Aphalo.

Figure 2.24: Filters in the field, stretched between two metal rods. By rolling the film onto the rods, it can be kept under tension. The researchers are standing on wooden catwalks used for access to the plots. Ushuaia, Argentina. Photograph: Kevin Maloney.

Figure 2.25: Filter on a frame. The filter has narrow strips of one type of film on top of a base film of another material. Viikki campus, University of Helsinki, Finland. Photograph: Pedro J. Aphalo.

Figure 2.26: Filters mounted on chicken wire net on tree branches. The underside of the branches is open to allow air circulation to prevent overheating. Note the strings that restrain the branch to prevent damage in windy weather. Kuohu, near Jyväskylä, Finland. Photograph: Titta Kotilainen.

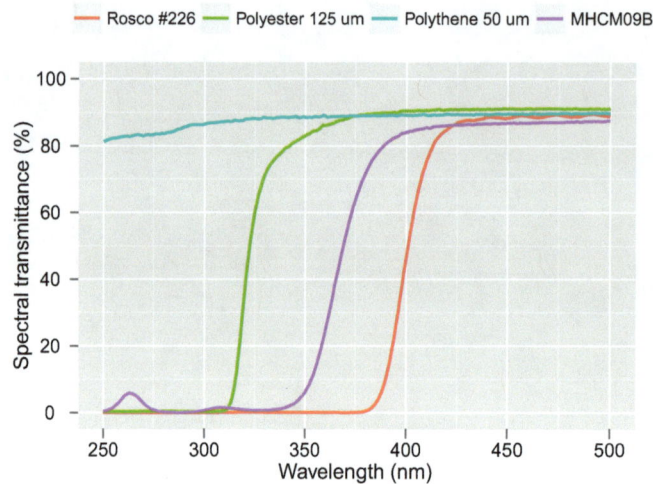

Figure 2.27: Transmittance spectra of several types of film used for altering the UV environment of plants. Rosco #226 is similar to Lee #226 and Formatt #226 films. MHCM09B is an experimental greenhouse film from BPI.Visqueen. The polyester film shown is Autostat CT5. The polythene is a generic product from Etola Oy (Finland) and 50 μm thick. Measurements were performed with an spectrophotometer equipped with an integrating sphere.

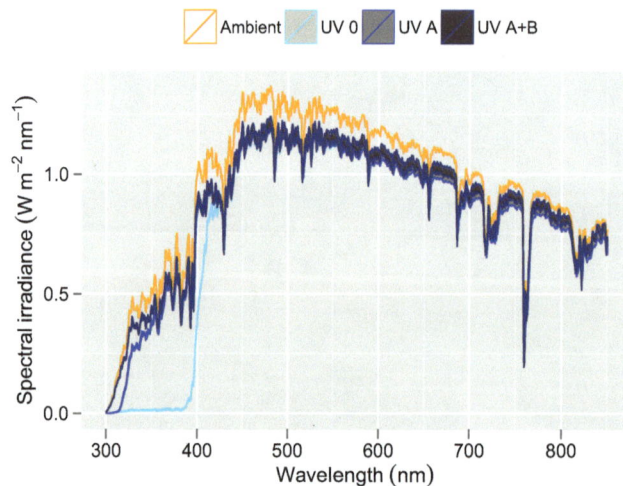

Figure 2.28: Spectral irradiance under of several types of film used for altering the UV environment of plants. UV 0: Rosco #226 is similar to Lee #226 and Formatt #226 films. UV A: Polyester film (Autostat CT5, similar to Mylar). UV A+B: polythene (generic product from Etola Oy, Finland, 50 μm thick). Measurements were performed with a Maya 2000 Pro (Ocean Optics) with stray light, and slit function corrections. Measurements done during May 2012, in Helsinki, Finland under clear sky conditions, at noon. The lines give the means from measurements under four replicate filters of each type, the grey band indicates ±1 s.e. Unpublished data of T. Matthew Robson and Saara Hartikainen.

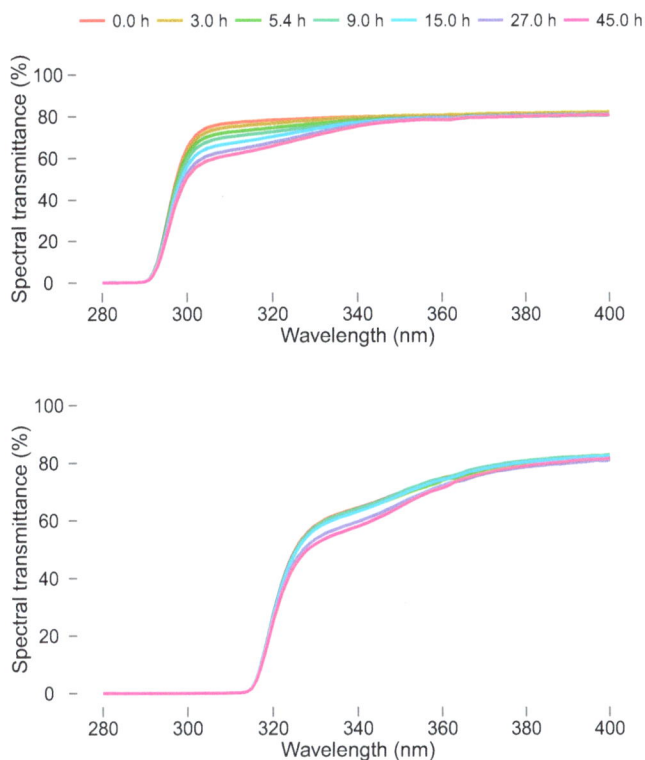

Figure 2.29: Transmittance spectra of 115 µm thick cellulose diacetate (top) and 125 µm thick polyester (bottom) films exposed to radiation from Q-Panel UVB-313 lamps at 25 cm distance in a greenhouse. Total time in the greenhouse was the same (9 d), but daily irradiation varied among the different samples. Measurements done with an array detector spectrophotometer not equipped with an integrating sphere. Data measured by Tania de la Rosa.

under and around the filters is a common problem which can lead to large unwanted changes in the microclimate in experimental plots.

The impracticality of filtering entire trees means that branch filters are often the most acceptable substitutes when assessing the effects of solar UV radiation on trees. However their installation and maintenance involves several additional considerations beyond those attached to fixed frames. Branch filters should be located only on the top and sides of branches, while the underside should be left open for ventilation and access. A structure of chicken wire and aluminium cables can maintain the shape of the filter while minimising contact between the filter and the treated part of the branch, however the weight of the structure should be kept to a minimum to reduce shading and the risk of mechanical damage to the tree. Filters mounted on branches behave like flags in the wind, so branches on which filters are installed need to be tethered to the ground or their movement stabilised using wooden canes attached to a different large branch

or limb of the tree, or preferably a combination of both these restraints, so that they do not move un-naturally or break in windy weather. By necessity in such experiments, the filtered area is relatively small and requires frequent maintenance to keep the shoots and leaves from growing into the filter or growing too far away from the filter into areas where they receive too much unfiltered and diffuse UV radiation. Particular attention should be given to the orientation of branch filters on a tree. For instance, branches and filters orientated to the north receive a very different dose of visible and UV radiation from those orientated to the south (Rousseaux et al., 2004). Comparing similar branches of the same tree under different filtration treatments is an obvious way to reduce the random variability among experimental units, but it is important to be aware that signals could potentially pass between different filtered branches and unfiltered branches which may dilute the response of a particular branch to its UV treatment. One method of controlling for communication between branches involves excising the phloem of

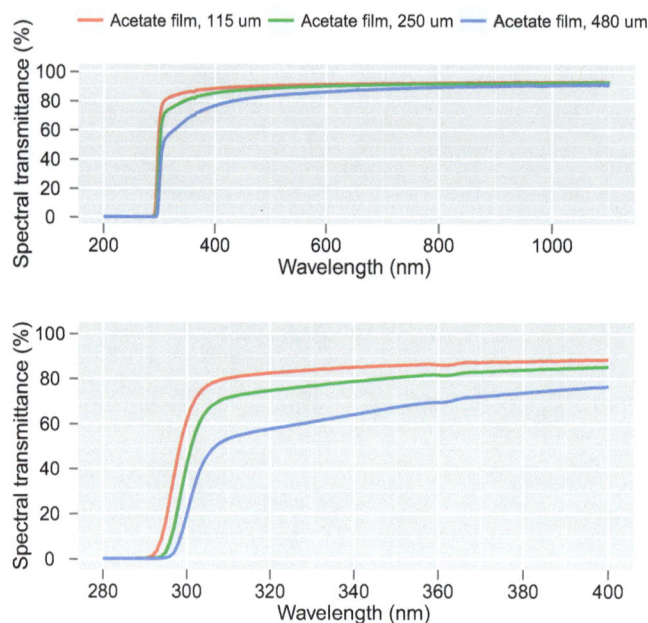

Figure 2.30: Transmittance spectra of new cellulose diacetate films of different thickness. Measurements made with an array detector spectrophotometer not equipped with an integrating sphere.

filtered branches. However, this isolation brings other problems as the realism of the experiment is reduced, so the need to adopt this approach will depend on the aims of each individual experiment. Maintaining branch filters in situ during the autumn and winter is not always practically feasible, however when using species that have determinate growth, their environment, particularly the radiation they receive, during bud set and the formation of leaf primordia can influence subsequent growth and leaf traits during the following growing season.

Sometimes the design of an experiment requires intermediate levels of UV attenuation, and this can be achieved by mounting strips of one type of plastic film on a base of another type (Figure 2.25). In such a system the strips are usually fastened at the edges of the wooden or wire frame on which the whole filter is attached. Unless the frames are small (less than 30 to 50 cm long sides) it may also be necessary to attach the strips in the middle with tape. In such a case the tape used must be clear to both PAR and UV radiation. The cheapest old-fashioned stationery tapes use a cellulose acetate substrate and are good for this purpose (e.g. Scotch Crystal Clear). Strips are usually 10 to 15 mm wide and if the desired effect is 50% attenuation, they are attached with a separation equal to their own width. Cutting the strips with scissors is a big task, but printers' shops can usually cut them effi-

ciently for a small charge. The UV irradiance under filters with strips is heterogeneous but the patches move with solar transit through the sky. This is more of a problem for measurement of irradiances than for the application of the treatments as reciprocity between irradiation time and irradiance apparently holds for such experiments (de la Rosa et al., 2001).

Even if we use filters that are totally opaque to UV-B, the treatment will not lead to total UV-B exclusion. That is why we use the phrase 'UV-B attenuation' for such treatments (Figure 2.31). As discussed in section 1.4 on page 8, solar UV radiation includes a large proportion of diffuse radiation coming from the sky, even under non-cloudy conditions. Consequently, some UV radiation penetrates under the edges of filters. It is therefore necessary to avoid locating experimental plants near the edges of the filters, and we recommend to have at least one row of border plants, forming a so-called 'guard row', from which no data are collected. Furthermore, it is possible to use curtains of the same filter material on the sides of the frames, but one should not close all sides as air movement is needed to avoid elevated temperatures and unnaturally still air. The filters should not be too high above the plants and irradiance should be measured at all the different positions where measured plants are located, rather than just under the centre of the frame.

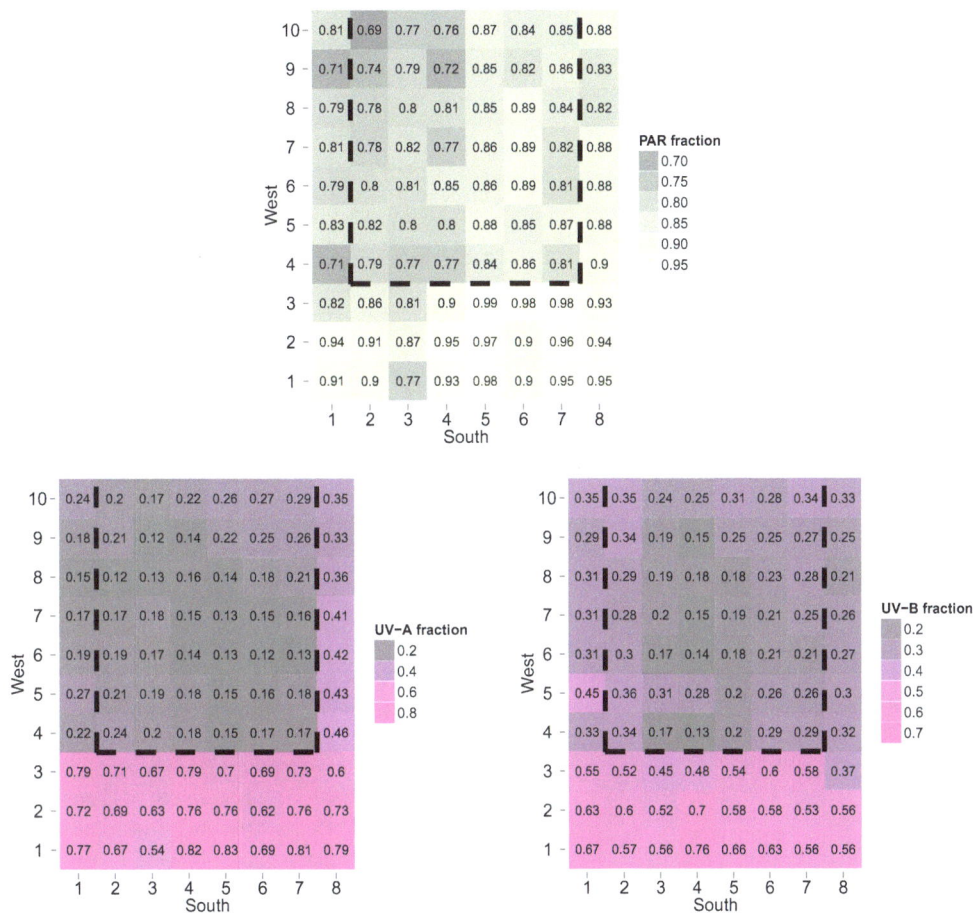

Figure 2.31: Maps of the relative photon irradiance under filter frames. (Top) PAR 40 cm below filters as a fraction of PAR photon irradiance above filters (average of polythene, polyester, and Rosco #226). (Bottom left) UV-A radiation 40 cm below Rosco #226 filters as a fraction of UV-A photon irradiance above filters. (Bottom right) UV-B radiation 40 cm below polyester filters as a fraction of UV-B photon irradiance above filters. Filters were mounted on 100×80 cm wooden frames. An average is given of measurements on clear-sky days under 4 replicate filters between 11:30 and 13:30 during May 2012 in Helsinki, Finland. The solid black line represents the edge of the shadow of the filter (due to the solar angle) and unfiltered solar radiation also partially fell on the areas outside the dashed lines. Each square in the map is 10×10 cm. The area outside the dashed line received unfiltered sunlight at least during some of the measurements. Measurements were done with an array detector spectroradiometer consisting of a Maya 2000 Pro spectrometer (Ocean Optics, Dunedin, Florida, USA) plus a D7-H-SMA cosine diffuser (Bentham, Reading, U.K.). Unpublished data of T. Matthew Robson, Saara Hartikainen, and Oriane Loiseau.

2.3.3 Measuring the spectral transmittance of filters

The recommended instrument for measuring filter spectral transmittance is a spectrometer or spectrophotometer equipped with an integrating sphere. Using an integrating sphere ensures that all transmitted radiation, both non-scattered and scattered is taken into account in the measurement. For non-scattering filters, one can obtain a reasonable approximation of the total spectral transmittance with a regular spectrophotometer. In most cases it is adequate to cut a piece of the film to the size of a cuvette and insert it carefully in the cuvette holder. In some models with an open light path (e.g. HP 8453, Hewlett Packard Gmbh, Waldbronn, Germany, now Agilent) it is possible to put a larger piece of filter material in the light path outside the cuvette holder but close to it. For measuring in the UV-B band one needs an instrument using a deuterium lamp as radiation source. The spectral resolution is important, and ideally should be 1 or 2 nm.

The optical properties of all filters change with time, even if the material from which they are made is stable, because filters get scratched and dirty. Dust and pollen stick to the filter surfaces and particles attach more tenaciously to some materials than others so that the filters in some treatments may become dirtier than in others. In addition to cleaning and replacing the filters regularly, it is necessary to measure the spectral transmittance of both new and used filters. We also recommend that samples of the filters are taken both at the start and end of their use and are stored protected from light and heat at least until the results of the experiments have been published, but preferably for longer.

2.4 Manipulating UV-B in the aquatic environment

2.4.1 Incubations in the field

In contrast to the situation in terrestrial environments, field experimentation with aquatic organisms is constrained not only by the underwater light conditions but also by a suite of physical perturbations (e.g. wave action, tidal fluctuations, currents, sedimentation, etc.). In the case of benthic macrophytes, an important factor that has to be considered is their vertical distribution. Deep or constantly submerged environments impose different difficulties compared to e.g. intertidal locations, which are subjected to changing light conditions due to the action of tides. In lakes and rivers, differences in the concentration of suspended UV absorbing substances between the surface and deeper habitats can be important. An experimental design for the study of seaweeds

normally includes different filters cutting-off UV-B and UV-A (see section 2.3 for details).

Algae can be exposed to these different radiation conditions by using self-made incubators constructed with plastic frames and nets with different diameters and forms depending of the size of the macrophyte (Figure 2.32). Both the cylindrical and sheet Plexiglass incubators can be covered by appropriate UV cut-off films and located at different water depths by using ropes or buoys. E.g. as used by Gómez et al. (2005) with the red alga *Gracilaria chilensis* in the Quempillen Estuary (Chile). The cylindrical incubators can be also used in intertidal pools, allowing them to float or locating them at the bottom of a pool. In general, experiments dealing with effects of natural UV radiation on aquatic organisms are carried out in a range of depths between 0 and 10 m to represent the UV penetration depth in coastal waters.

The diurnal variability of solar radiation is one of the most important factors to consider when working with aquatic macrophytes in the field. In general, at temperate and cold-temperate locations, daily changes in the irradiance conditions are also exacerbated by depth, which affects the number of hours during which algae are exposed to UV radiation. It is well known that photosynthetic physiology changes during the day with maximum rates of photoinhibition (a mechanism of dissipation of excess energy) occurring during the highest solar irradiance at midday, while a recovery in the photosynthetic capacity is found in the afternoon (Figueroa et al., 1997; Gómez et al., 2004; Häder et al., 1996). This pattern is well generalized in macrophytes and strongly dependent on the growth depth: individuals growing at shallow depths show a more efficient photoinhibition than their deeper counterparts (Franklin and Forster, 1997). Thus, experimental designs employing transplantation or incubations at different depths along the water column should consider these potential confounding effects.

Tides are also a relevant factor affecting UV radiation in shallow marine waters, especially for intertidal species. Fluctuations in the height of the water column caused by tidal regimes can reduce or increase considerably the available UV radiation. For example, in a study comparing two coastal systems with different tidal range, Huovinen and Gómez (2011) demonstrated that algae from fjords, with tidal ranges close to 7 m, exhibited less difference in photosynthetic light demand and susceptibility to UV radiation than algae from an open coast where maximal tidal variation was 2 m. Probably, this is a major factor influencing experimentation with benthic intertidal macrophytes, especially in highly dynamic systems that may also be characterized by strong wave action.

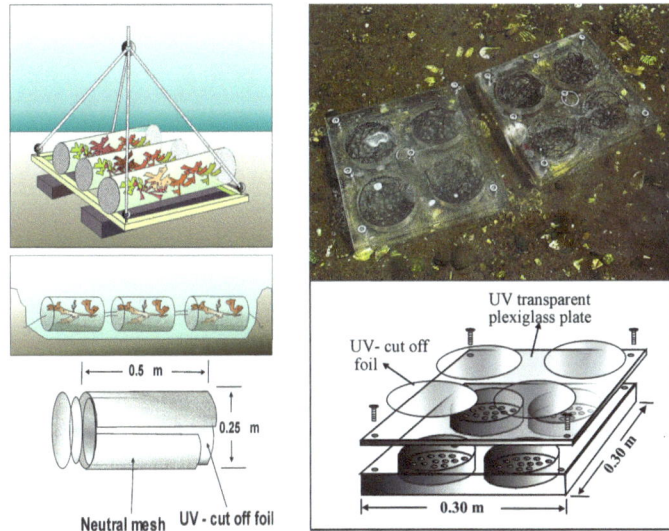

Figure 2.32: Experimental setup used for incubation of marine macrophytes to underwater UV fields.

2.4.2 Incubation under artificial lamps

The effect of UV radiation on aquatic plants can also be studied using incubations under controlled illumination conditions (using lamps) in the laboratory (Figure 2.33). Normally the UV sources are similar to those used for incubations of terrestrial plants (see section 2.2.1).

In general, experimentation on aquatic plants using small vessels under laboratory conditions is designed to test for effects of UV in isolated cells or small sized macrophytes. In the case of large seaweeds (e.g. kelps), space limitation restricts incubations to pieces or sample discs. In the case of seaweeds, due to their simple morphological organization, the use of cut sections still allows a realistic extrapolation of the situation in the whole thallus.

The optical characteristics and shape of the vessels containing the samples strongly determine the orientation of the sample to the UV source and the experimental setup. Often the vessels used in routine incubations are opaque to UV radiation (e.g. Pyrex, and many types of glass ware and polycarbonate), and thus lamps should be located above them. However, there are various UV transparent materials, e.g. methacrylate (Type GS-2458) or quartz glass, which permits UV irradiation from different sides and mimics better the natural underwater conditions. In experimental setups which cover vessels with UV cut-off filters, bubbles from aeration can affect the UV penetration and diffraction patterns within the vessels. Thus, cut-off and neutral filters should be kept at sufficient distance from the water.

Due to the heat emission of lamps, temperature within the incubation chamber can vary considerably, which

should be taken into account, similar as with incubations of terrestrial plants. This factor is especially relevant for algae subject to small changes in temperature in their habitat, e.g. some deep water and polar algae. Thermoregulated water baths can minimize this problem (Figure 2.33).

The use of artificial UV sources presents various advantages. Firstly, the manipulation of the UV:PAR ratio can permit the examination of some processes that normally are masked by the prevailing high PAR irradiances in the field, as well as simulations of different UV scenarios. Secondly, it is possible to standardize conditions for physiology, a situation not normally possible in the natural habitat, where unpredictable environment conditions (e.g. water motion, scattered light field) make *in situ* experimentation difficult.

2.5 Suitable treatments and controls

In every experiment one must include a suitable control in addition to treatments. What is a suitable control will depend on the aims of the study. In order to assess the effect of UV-B radiation, then the only difference between an UV-B treatment and the control treatment should be the irradiance of UV-B. This may seem obvious, but it is important to think about the consequence of this for the design of experiments. Many UV-B sources emit in addition to UV-B radiation, UV-C, UV-A and visible radiation. Common UV-A sources emit in addition to UV-A, UV-C, UV-B and visible radiation. As most fluorescent and some other UV sources emit small amounts of UV-C radiation, which is absent in sunlight, UV-C radiation must be re-

Figure 2.33: Examples of incubation set up of aquatic macrophytes under UV lamps. Photograph: Iván Gómez.

moved by filtration. See Box 2.1 for descriptions of some typical experimental designs and their advantages and pitfalls.

When using filters, one should take into account the side-effects of using them. In addition to absorbing different amounts and wavelengths of UV radiation, filters may have effects on visible and infra-red irradiance. No filter transmits 100% of visible light. In addition filters block precipitation and wind, and may increase the temperature of the air, vegetation and soil below them. So, one should always use, as the control for assessing effects of UV attenuation, a UV transparent filter rather than no filter. See Box 2.2 for some examples of good and bad experimental setups.

In the case of controlled environments, various combinations of lamps and filters are used to create different UV radiation treatments. As discussed above, if in order to test for effects of UV-B radiation, UV-B exposure should be the only difference between treatment and control conditions. For example, one can have a sun simulator, or at least a combination of lamps providing both UV- and visible radiation and then have different filters between the lamps and plants. The same principles as discussed in Examples 2.1 and 2.2 apply to experiments in controlled experiments.

2.6 | Recommendations

2.6.1 Recommendations for outdoor experiments

In this section we list recommendations related to the manipulation of UV radiation in outdoor experiments, and in section 2.6.2 list some additional recommenda-

tions on UV radiation manipulations inside greenhouses and controlled environments. See section 3.15 for recommendations on how to quantify UV radiation, section 4.9 for recommendations about growing conditions, and section 5.12 for recommendations about statistical design of experiments.

1. Make sure that the only difference between the treatments you compare is what you want to test. If possible make measurements of the environmental conditions in the different treatments and report them in your publications.

   a) In experiments with UV-B lamps avoid differences in PAR irradiance among treatments. Such confounding differences can be caused by shade from lamps and the frames supporting them if they are kept at different heights. All treatments must have lamps that are switched on or off, covered with different filters or dimmed to different radiance values.

   b) In experiments with UV filters avoid difference in PAR irradiance and scattering. Filters used to block UV radiation and the frames supporting them have some attenuating and scattering effect on PAR. All treatments must have similar supporting structures, be positioned at the same height, and the filters themselves must have as similar as possible transmittance in regions of the spectrum outside the UV region. All filters must have as similar as possible light scattering properties, as even if PAR irradiance is the same, differences in the proportion of diffuse light affects the growth of plants.

Box 2.1: Treatments and controls when using lamps in the field

Case We use Q-Panel UVB-313 tubes, or Philips TL12 fluorescent tubes outdoors. These lamps are sold as broadband UV-B lamps, but they emit radiation of other wavelengths (UV-C, UV-A, and visible) in addition to UV-B (see Figure 2.2 on page 37). Our objective is to study the effect of UV-B radiation.

Design 1 We remove UV-C from the UV-B treatment by filtering the lamps with cellulose diacetate (see Figure 2.27 on page 58). We add UV-A (and a very small amount of visible radiation) to the controls by having identical lamps as in the treatments but filtered with polyester film, which absorbs UV-C and UV-B. This also ensures that any effect of the lamps on the temperature of plants is similar in treated and control plants, and shading of sunlight is similar. This type of control is usually called 'UV-A control'.

Design 2 As above, but in addition we add a second control with unenergized lamps. We can compare this control to the UV-A control to assess whether there is any side effect related to the functioning of the lamps but unrelated to UV-B radiation.

Design 3 We add a third control with no lamps or frames. By comparing this control with the control with unenergized lamps we can test for the effect of shading by lamps and supporting structures.

Design 4 We use cellulose diacetate-filtered UV-B lamps for treatments and unenergized lamps for controls. In this case the effect of UV-B is confounded with other effects of the UV-B lamps, except for shading.

Design 5 We use cellulose diacetate-filtered UV-B lamps for treatments and no lamps or frames for the controls. In this case the effect of UV-B is confounded with all other effects of using lamps.

Caveat When discussing very small effects of treatments and the different controls there is always the risk of misinterpretation as the films used as filters are far from perfect, there is a transition zone between high absorptance and high transmittance spanning tenths of nanometres in wavelength (see Figure 2.27). Consequently, UV-A controls will not be exposed to exactly the same irradiance of UV-A as UV-B treated plants. Usually they will be exposed to a slightly lower irradiance of UV-A plus a trace of UV-B.

Comparison Design 1 is the simplest design that can be used to test for the effects of UV-B. Design 2, is preferable as it allows an assessment of the possible secondary effects of lamps, except for shading. Design 3, is rarely used, but allows an assessment all side-effects of lamps in addition to the effect of UV-B. Is useful in an ecological context where we are interested in assessing how much the experimental setup disrupts natural conditions. Design 4 should be avoided as it is unsuitable for testing the effects of UV-B radiation as all the different effects of the lamps are confounded with the effect of UV-B enhancement, except for shade. Design 5 is the worst possible, and should never be used.

c) In experiments with UV-B lamps avoid differences in temperature among treatments. Keep the distance between UV lamps and the top of the plants at least 0.4–0.5 m. If enclosures like open-top chambers are used, provide enough ventilation or cooling to remove the heat generated by the lamps.

d) In experiments with UV filters avoid differences in temperature among treatments. The main factor to consider in this case is ventilation and distance from the film to the top of the plants. An additional factor is the transmittance of the different films to longwave infra-red radiation. The filters can affect the temperatures both during the day and at night. If possible, measure the temperature of the plants and soil, in addition to the temperature of the air.

e) In experiments with UV filters avoid artifacts caused by plasticisers used in some plastic films used as filters. Ensure good ventilation and if possible avoid using cellulose diacetate films for near-ambient controls. When not using lamps there is no need to remove UV-C radiation and consequently the more stable and less toxic PTFE (Teflon) or polythene films

---

Box 2.2: Treatments and controls when using filters in the field

Case We use polyester ('Mylar') film outdoors to attenuate UV-B radiation (see Figure 2.28 on page 58). Our objective is to study the effect of solar UV-B radiation.

Design 1 We attenuate UV-B radiation (but not UV-A radiation) in the −UV-B treatment by filtering sunlight with a polyester film. We use as a control a film that transmits UV-B and UV-A. This ensures that most of the effect of the films on precipitation, wind and temperature is similar for treated and control plants. Also shading of sunlight by supporting structures is similar. This type of control is usually called 'near ambient UV control'.

Design 2 As above, but in addition we add a second treatment with a filter absorbing both UV-A- and UV-B radiation. We can compare this treatment to the UV-B attenuation treatment to assess whether there is an effect of UV-A radiation.

Design 3 We add a second control without filters or frames ('ambient control'). By comparing this control with the near-ambient control we can test for the effect of filters and supporting structures.

Design 4 We attenuate UV-B radiation in our −UV-B treatment by filtering sunlight with a polyester film. We have as controls plots with no filters or supporting structures. In this case the effect of UV-B is confounded with other effects of the filters.

Caveat When discussing very small effects of treatments and the different controls there is always the risk of misinterpretation as the films used as filters are far from perfect, there is a transition zone between high absorptance and high transmittance spanning tenths of nanometres in wavelength (see Figure 2.27 on page 58). Consequently, near ambient UV controls will not be exposed to exactly the same irradiance of UV-A as plants under the UV-B-attenuation treated plants. Usually they will be exposed to a slightly lower irradiance of UV-A plus a trace of long wavelength UV-B.

Comparison Design 1 is the simplest design that can be used to test for the effects of UV-B. Design 2, is preferable as it allows us to also assess the effects of solar UV-A radiation. Design 3, is not always used, but allows to assess all side-effects of filters in addition to the effect of UV radiation. This design is useful in an ecological context where we are interested in assessing how much our experimental setup disrupts natural conditions. Design 4 should be avoided as it is unsuitable for testing the effects of UV-B radiation as all the different effects of the filters are confounded with the effect of UV-B attenuation.

---

should be used instead.

2. In experiments with UV-B lamps do not use unfiltered lamps. Most UV-B lamps also emit some UV-C that must be filtered with a cellulose diacetate filter in UV-B treated plots. Most UV-B lamps also emit UV-A, so an additional UV-A control should be included in all experiments. This control is achieved by filtering out both UV-C and UV-B radiation by means of polyester film. The smallest well designed experiment should include three treatments: (i) a control with unenergized lamps, (ii) an UV-A control with polyester filtered lamps, and (iii) an UV-B treatment with cellulose diacetate filtered lamps.

3. If the difference between near-ambient controls and true ambient conditions is of interest, then an additional true ambient control should be included to test how big is the effect of the small differences in air, soil and leaf temperatures, PAR irradiance, air humidity and any other side-effect of the manipulations. In any case, when designing an experiment we should strive to minimize shading, alterations in temperature and ventilation.

4. In experiments with UV-B lamps, if at all possible, use modulated systems that avoid unrealistic ratios between UV- and PAR irradiances. If the use of a modulated system is impossible, keep the lamps on only a few hours centred on solar noon and switch them off during cloudy weather.

5. In experiments with UV-B lamps use high frequency electronic ballasts rather than electromagnetic ballasts to drive the lamps, so as to avoid flicker in the UV radiation output, which can affect insect behaviour.

6. In all experiments using filters check periodically whether the spectral characteristics of the filters have changed and replace them when needed. Cellulose diacetate degrades particularly fast and, for example when used at about 0.3 m from UV-B lamps it should be replace after about 50 h of lamp irradiation. If wrapped on lamps, it should be replaced even more frequently. These times are approximate and can also be extended when using modulated systems.

7. In experiments using filters to attenuate UV in sunlight, even if the filters are not yet degraded or if they are made of stable materials like glass or PFTE (Teflon) the transmission characteristics will change by the accumulation of dust, dirt and pollen. If this happens, the filters should be cleaned or replaced.

2.6.2 Recommendations for experiments in greenhouses and controlled environments

Many of the recommendations in section 2.6.1 can be adapted to apply to indoor experiments. Here we list additional recommendations applicable to experiments in greenhouses and controlled experiments.

1. Keep the balance between UV-B irradiance, UV-A irradiance and PAR as similar as possible to that in sunlight and/or vegetation canopies. Many growth chambers and growth rooms achieve relatively low irradiances of PAR. Avoid using high doses of UV-B in such cabinets. Some chambers and rooms have sheets of polycarbonate (PC) separating the lamps from the plants, in such chambers UV levels are negligible. In chambers using bare lamps, UV-A irradiance is unrealistically low, but usually not equal to zero. A visible- and UV radiation spectral composition truly matching sunlight can be achieved only with sun simulators.

2. Unless you are specifically studying the effect of step changes in UV-B irradiance, for annual plants we recommend that the UV-B treatment starts at the time of seed germination or earlier, and for perennial plants before budburst.

3. In greenhouses, depending on the cladding material used, the UV-B and UV-A irradiances will differ from those outdoors. Irradiances of PAR and UV are always somewhat lower than outdoors as the cladding materials and the structure of the greenhouse absorb part of the radiation.

4. Be aware that even though the temperature may be adequate for growing plants in a greenhouse during the winter, light and especially UV irradiances are much lower than in the summer, even when high pressure sodium or metal halide lamps are used to increase PAR levels. For this reason, in experiments with UV-B lamps the balance between PAR and UV can become very different to that under natural conditions.

5. Of course, if you are not interested in what happens under natural conditions but are researching the management of crops under cover, then you only have to make sure that your treatments match what can be achieved in commercial production systems.

2.7 Further reading

The following links are to directives and recommendations concerning protection from UV exposure.
http://eur-lex.europa.eu/LexUriServ/
LexUriServ.do?uri=OJ:L:2006:114:0038:
0059:EN:PDF (EU directive 2006/25/EC about protecting workers from optical radiation)
http://www.hse.gov.uk/radiation/
nonionising/optical.htm (guide on how to apply this directive (not approved yet by EU)
http://www.who.int/uv/sun_protection/
en/ (general recommendations by WHO)

2.8 Appendix: Calculating the radiation field under an array of lamps

This section includes code for calculating the radiation field under different arrays of lamps (Figures 2.5 and 2.6). The algorithm is that in the BASIC program in Björn and Teramura (1993) with a few changes related to the density of grid points used, and the area of the grid for which the light field is calculated. The grid area is expanded as it is also important to assess the distance at which neighbouring arrays can be located. Please, see Björn and Teramura (1993) for the details of the algorithm. The algorithm assumes that there are no reflecting surfaces near the array, so it is better suited for simulating the light field under arrays of lamps located outdoors, than for those located in controlled environments.

The functions are in the R package 'lamps' which is available from the handbook web site. The following example also uses package ggplot2, available from CRAN (comprehensive R archive network, at `http://cran.r-project.org/` and many mirror sites). The first example (Box 2.3) is for calculating and plotting the radiation field under and array of fluorescent tubes, using mostly the defaults. See the package documentation for the many function arguments available. The second example shows how to draw the positions of the fluorescent tubes (Box 2.4).

2.9 Appendix: Suppliers of light sources and filters

In this section we provide names and web addresses to some suppliers of light sources and filters. This is certainly an incomplete list and exclusion reflects only our ignorance.

UV lamps:

`http://www.lighting.philips.com/` (UV-B and many other types of lamps)

`http://www.q-lab.com/` ('*Q-Panel*' UV-A and UV-B lamps)

Ballasts for fluorescent lamps

`http://www.osram.com/` ('*Quicktronic*' electronic dimming ballasts)

`http://www.tridonic.com/` (electronic dimming ballasts)

LEDs:

`http://www.osram.com/`

`http://www.lighting.philips.com/`

`http://www.roithner-laser.com/`

`http://www.valoya.com/`

Xenon arc lamps:

`http://www.newport.com/oriel/`

`http://www.muller-elektronik-optik.de/`

Lasers:

`http://www.opotek.com` (Tunnable UV lasers)

`http://www.roithner-laser.com/`

Filters:

`http://www.formatt.co.uk/` (theatrical 'gels')

`http://www.leefilters.com/` (theatrical 'gels')

`http://www.macdermidautotype.com/` ('*Autostat*' polyester film)

`http://www.nordbergstekniska.se` (cellulose diacetate, Mylar and other films)

`http://www.rosco.com/` (theatrical 'gels')

`http://www.schott.com/` (glass filters, and special glasses)

`http://www.thermoplast.fi/` ('*Autostat*', '*Aclar*', and many other types of films.

Box 2.3: Calculating and plotting the radiation field under an array of fluorescent tubes with R, using packages 'lamps' and 'ggplot2'.

```r
library(lamps)
library(ggplot2)
# cosine arrangement
light_field45.data <- light_field()  # default for z is 0.45
light_field60.data <- light_field(z = 0.6)
light_field30.data <- light_field(z = 0.3)
# equidistant
light_field45_l.data <- light_field(cosine_dist = FALSE)
light_field60_l.data <- light_field(z = 0.6, cosine_dist = FALSE)
light_field30_l.data <- light_field(z = 0.3, cosine_dist = FALSE)
# a simple figure
fig1 <- ggplot(light_field45.data, aes(x = x, y = y, fill = c))
fig1 <- fig1 + geom_tile() + geom_contour(aes(z = c), colour = "white",
    binwidth = 0.1)
fig1 <- fig1 + labs(x = "x (m)", y = "y (m)", fill = "relative\nirradiance")
print(fig1)
# facets, figure with six panels we prepare the data by row
# binding
len <- length(light_field45.data$x)
irrad_all.data <- rbind(light_field30.data, light_field45.data,
    light_field60.data, light_field30_l.data, light_field45_l.data,
    light_field60_l.data)
irrad_all.data$dist <- c(rep("'cosine' pattern", len * 3), rep("equidistant",
    len * 3))
irrad_all.data$distance <- rep(rep(c("z = 0.30 m", "z = 0.45 m",
    "z = 0.60 m"), rep(len, 3)), 2)
# we draw the figure
fig <- ggplot(irrad_all.data, aes(x = x, y = y, fill = c))
fig <- fig + geom_tile() + geom_contour(aes(z = c), colour = "white",
    binwidth = 0.1)
fig <- fig + labs(x = "x (m)", y = "y (m)", fill = "relat.\nirrad.")
fig <- fig + facet_grid(distance ~ dist) + coord_equal(ratio = 1)
print(fig)
```

69

Box 2.4: Calculating and plotting the lamp positions in an array of fluorescent tubes with R, using packages 'lamps' and 'ggplot2'.

```r
library(lamps)
library(ggplot2)
# calculating the positions of the tubes in the array
positions_cos.data <- tube_positions()
positions_eq.data <- tube_positions(cosine_dist = FALSE)
# row binding the data
positions.data <- rbind(positions_cos.data, positions_eq.data)
# adding labels
positions.data$dist <- rep(c("'cosine' pattern", "equidistant"),
    rep(length(positions_cos.data$x), 2))
positions.data$fake_label <- factor(rep("lamps", length(positions.data$dist)))
```

```r
# drawing the figure
fig_x <- ggplot(data = positions.data, aes(x = x, y = y, colour = tube))
fig_x <- fig_x + geom_line(size = 2) + coord_equal(ratio = 1)
fig_x <- fig_x + facet_grid(fake_label ~ dist)
fig_x <- fig_x + labs(x = "x (m)", y = "y (m)", colour = "tubes   ")
fig_x <- fig_x + ylim(c(-1.48/2, 1.48/2)) + xlim(c(-1.1, 1.1))
print(fig_x)
```

3 Quantifying UV radiation

Lars Olof Björn, Andy McLeod, Pedro J. Aphalo, Andreas Albert, Anders V. Lindfors, Anu Heikkilä, Predrag Kolarž, Lasse Ylianttila, Gaetano Zipoli, Daniele Grifoni, Pirjo Huovinen, Iván Gómez, Félix López Figueroa

3.1 Basic concepts and terminology

3.1.1 Introduction to UV and visible radiation

In this section we will discuss the quantities, units and terminology used to describe visible and UV radiation, and explain how to use and interpret information about radiation. We will explain the basic concepts associated with quantifying radiation and how the different aspects of this topic are applied to UV experimentation.

When describing experimental conditions we need to avoid all ambiguity, so that our results can be interpreted and experiments repeated. For this reason "Light intensity" and "amount of light", which are ambiguous terms, should be avoided in scientific contexts unless it is made absolutely clear what they stand for.

Radiation always consists of different wavelength components, although we often use the term "monochromatic" when the spectral range is narrow. Not even a single photon can be assigned an exact wavelength. As radiation also travels in various directions, it is also always associated with a distribution of directions. When we say "collimated radiation", we mean that the angular distribution is very narrow. When we say "scattered or 'diffuse' radiation", we mean that the angular distribution is wide. In order to fully describe radiation, we must therefore describe the distribution of its wavelength components (the spectral distribution) and their direction in addition to "amount". Polarization is another property of radiation, but it will not be dealt with here. Finally, we should consider the time dimension; radiation changes over time. Radiation can be measured almost instantaneously and expressed as a rate, or integrated over time. Consequently we must distinguish between total energy, or the total number (or total number of moles) of photons, and the energy-or-photons per unit time. The first case gives an accumulated quantity, and the second case gives a rate. A clear distinction should also be made between the incident radiation on a target and the amount of radiation absorbed. When dealing with ionizing radiation, the term "dose" is always used to mean the radiation absorbed, however the same term often designates incident radiation when used in connection with ultraviolet-B and visible radiation.

A comprehensive glossary of terms relating to visible and ultraviolet radiation has been published by Braslavsky (2007). It can be downloaded from the Internet (see link in reference list).

3.1.2 Direction

Daylight radiation has two components: (a) radiation arriving directly from the sun, or direct radiation, and (b) radiation arriving from the sky, or scattered or 'diffuse' radiation. Because it is very distant compared to its size, the sun behaves almost as a point source. For this reason direct radiation at the Earth surface comes from a single direction, so that its 'rays' are parallel or collimated. Diffuse radiation is due to scattering and reflection in the atmosphere, and arrives from the whole sky, its 'rays' are not parallel. Diffuse radiation is not collimated.

Light meters and radiation sensors are usually calibrated using radiation from (approximately) a single direction, i.e., collimated radiation from a lamp, but in nature (where most plants live!) radiation is not collimated. The solar radiation reaching the ground on clear cloud-free days as direct sunlight is rather well collimated, but in addition to direct radiation there is diffuse radiation from the sky. Furthermore, radiation reflected by the ground and objects in the surroundings also contributes to the diffuse radiation received by plants. A plant scientist wishing to understand how plants use and react to radiation has to take both direct and diffuse radiation into account.

Many radiation meters have a flat receiving surface,

just like many plant leaves, and so should in principle be well suited to measuring the radiation reaching a leaf. But if we are interested in the whole plant rather than a single leaf the situation is different, since plants are three-dimensional not flat and different surfaces on the plant face in different directions. In such situations where we wish to obtain an approximate estimate of the quantity of radiation incident on a plant, it is better to use a sphere than a flat radiation meter. This is achieved with a measuring instrument that is equally sensitive to radiation from any direction.

This brings us to the distinction between

1. Irradiance, i.e., radiation power incident on a flat surface of unit area, and

2. Energy fluence rate (or fluence rate for short), i.e., radiation power incident on a sphere of unit cross section. The term fluence rate was introduced by Rupert (1974).

Other terms with the same meaning as energy fluence rate are space irradiance, scalar irradiance and actinic flux. The latter is used mostly by atmospheric scientists. The term spherical irradiance has been used in similar contexts, but means one quarter of the fluence rate. Vectorial irradiance is just the same as irradiance.

Both irradiance and energy fluence rate, can also be described in terms of photons. For item 1 above, the corresponding photon nomenclature has not yet been standardised. It would be logical to use the term photon irradiance, but many people, especially in the photosynthesis field, use the term photon flux density and the abbreviation PFD (PPFD for photosynthetic photon flux density). For item 2 the term photon fluence rate is well accepted among plant physiologists, but hardly among scientists in general (for example meteorologists call it 'actinic flux' or 'scalar irradiance'). Energy fluence is the energy fluence rate integrated over time. The word energy is frequently omitted, so that 'fluence' has the same meaning as 'energy fluence'. In contrast, when 'photon fluence' is meant, the word 'photon' is always explicitly mentioned.

When giving a value for irradiance, the direction of the plane for which the irradiance is considered must be specified. This is often the horizontal plane but will depend on the object studied: for instance, the surface a leaf is frequently not horizontal. For collimated radiation (coming from a single direction), irradiance and fluence rate have the same numerical value if the beam of radiation is perpendicular to the plane on which irradiance is measured (Figure 3.1.A). For completely isotropic radiation coming from above (equally from any direction above the horizontal) the fluence rate is twice the irradiance on a horizontal plane (Figure 3.1.C). Sometimes one may

also have to consider radiation from below, especially in aquatic environments (Figure 3.1.D).

When a delimited beam of radiation hits a plane surface perpendicularly, it results in the irradiance, E, but the same beam when tilted at an angle α to the vertical will be more spread out, and thus the irradiance will be lower. More specifically, the irradiance will be $E \cdot \cos \alpha$ (Figure 3.1.B).

3.1.3 Spectral irradiance

When we deal with *how a quantity of radiation varies with wavelength*, for instance when plotting a spectrum, we add "spectral" to the name of the quantity to give "spectral irradiance" and "spectral fluence rate". We must also adjust the units to account for the width of the part of the spectrum that is included in the value specified (in photobiology nm^{-1} is often used to indicate wavelength interval, but the scale can also be defined in terms of wave number or frequency, as in, cm^{-1} or s^{-1} respectively).

3.1.4 Wavelength

It is impractical to always quantify radiation by giving its complete spectral composition; so simplifications have to be made that account for the most important features of a spectrum. From a purely physical viewpoint, there are two basic ways of quantifying radiation. Either we express the quantity related to the number of photons, or the quantity related to the energy of radiation. When considering radiation of a single wavelength, the energy of a photon, or quantum (in a vacuum), is inversely proportional to the wavelength and the proportionality constant is Planck's constant multiplied by the speed of light in a vacuum (see section 1.3).

$$q(\lambda) = h \cdot v = h \cdot \frac{c}{\lambda} \qquad (3.1)$$

where $q(\lambda)$ is a quantum, or the amount of energy that one photon has, h is Planck's constant ($h = 6.626 \times 10^{-34}$ Js), v is frequency, λ is wavelength, and c is the speed of light in a vacuum ($c = 2.998 \times 10^8$ m s^{-1}). Please, see section 1.4 on page 8 for a numerical example. When dealing with a band of wavelengths it is necessary to repeat these calculations for each wavelength (in practice a very narrow band) and only then integrate across wavelengths. For example, PAR in units of W m^{-2} is obtained by integrating the spectral irradiance from 400 to 700 nm. As explained in Box 1.2 on page 7, to determine a quantity in terms of photons, an energetic quantity has to be weighted by the number of photons, i.e. divided by the energy of a single photon at each wavelength from equation 3.1. To get moles of photons, the value has to be divided further by Avogadro's number ($N_A = 6.022 \times 10^{23}$

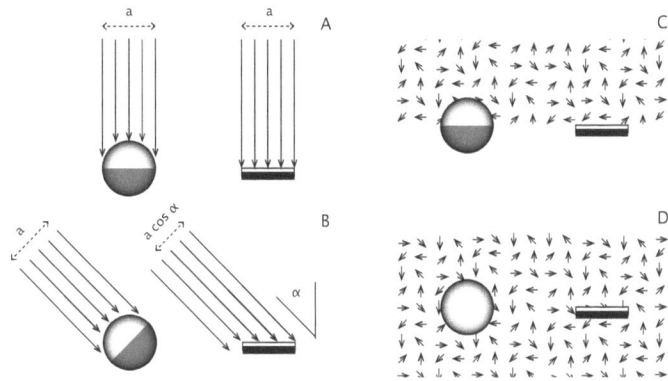

Figure 3.1: The concepts of irradiance and fluence rate explained. (A) A beam of radiation perpendicular to the plane of an irradiance sensor; irradiance and fluence rate have the same numerical value. (B) Radiation beam at an angle to the irradiance sensor; irradiance has a smaller numerical value than fluence rate. (C) Perfectly diffuse radiation from the hemisphere above the irradiance sensor; the numerical value of irradiance is one half that of fluence rate. (D) Perfectly diffuse radiation from both hemispheres, both above and below the sensors; the numerical value of irradiance is one quarter that of fluence rate. A spherical sensor (left) measures fluence rate, and a one-sided planar sensor (right) measures irradiance. In (A) and (B) radiation is collimated and represented by parallel arrows. In (C) and (D) radiation is diffuse and represented by "randomly" oriented arrows. In (A), (B) and (C) the shaded hemisphere of the spherical sensor is grey.

mol^{-1}). For example, the PAR photon irradiance or PPFD in units of $\mu mol\,m^{-2}\,s^{-1}$ is obtained by

$$PPFD = \frac{1}{N_A} \int_{400\,nm}^{700\,nm} \frac{\lambda}{hc}\, E(\lambda)\, d\lambda \qquad (3.2)$$

If we have measured (energy) irradiance, and want to convert this value to photon irradiance, the conversion will be possible only if we have information about the spectral composition of the radiation we measured. For PAR, 1 $W\,m^{-2}$ of "average daylight" (400–700 nm) is approximately 4.6 $\mu mol\,m^{-2}\,s^{-1}$. This figure is exact only if the radiation is equal from 400 to 700 nm, because it represents the value at the central wavelength of 550 nm. But to make the exact conversion, we must know the spectrum of irradiance measured at the time of measurement, as the solar spectrum varies through the day and with the seasons. In the UV-B band this is even more important as the UV-B tail of the solar spectrum varies much more than the PAR band. When dealing with lamps, we can sometimes use conversion factors obtained for a particular sensor and lamp combination to convert measured values from energy to photon irradiance. If the lamps are filtered, for example with cellulose diacetate, the spectrum transmitted by the filter will change as the filter ages, and consequently each time irradiance beneath the filter is measured the spectrum of the radiation source (i.e. the lamps) must also be measured.

3.1.5 Units used for photons and energy

Radiation considered as photons can be expressed either as number of photons, or as moles of photons (the symbol for moles is mol). An obsolete term for a mole of photons is an Einstein (E); it should no longer be used. According to Avogadro's number (N_A), one mole is 6.02217×10^{23} photons, but in most cases a more convenient unit is μmol (micromole of photons, 6.02217×10^{17} photons). Either of these units can be expressed per time and per area or (rarely in biological contexts) per volume.

The unit of energy is a joule (J). Energy per time is power, and a joule per second is a watt (W). Both can be expressed per area (or, rarely in biological contexts, per volume, i.e., energy density or power density). Simply giving a value followed by "$W\,m^{-2}$" without further qualification is not meaningful and should be avoided, since this leaves the surface you are expressing with m^2 undefined. Is it a flat surface or a curved one? If it is flat, what is its orientation? When reporting experimental conditions, all these factors have to be specified.

3.1.6 Ratios of UV and visible radiation

As already mentioned in section 1.1 on page 1, it is very important to conduct experiments under realistic environmental conditions. For the control of these experimental conditions, it is helpful—besides using biologically effective exposures—to calculate ratios UV:PAR and UV-B:UV-A:PAR. The ratios can also be used for comparison of different experimental conditions or experimental conditions with natural conditions outside in the field, as for example shown in section 2.2.7 on page 48 and in table 3.8 on page 109. For the calculation of these ratios, it is absolutely essential to use the same quantity of radiation or the same weighting procedure, respectively, i.e. energetic units or photon units, but not a mixture.

The term "energetic" means that the spectral irradiance is simply integrated over the respective spectral band yielding units like W m^{-2}. For the calculation of photon ratios it is also necessary to integrate over the respective spectral bands, but additionally weighted by the number of photons at each wavelength (see section 3.1.4). A function to calculate photon ratios in R is described in section 3.19.4 on page 116.

3.1.7 A practical example

The importance of indicating the type of sensor and its orientation is exemplified by the changing ratio of fluence rate to irradiance throughout a day (Figure 3.2). At noon, when the sun is high in the sky the ratio is at its minimum. Even then the ratio remains larger than one because, other than at the equator, the midday sun never reaches the zenith and because the sensor will also always receive diffuse radiation from the whole sky. We can say that when measuring solar radiation the fluence rate will always be numerically larger than the irradiance. Towards both ends of the day the ratio reaches its maximum value because irradiance is being measured on a horizontal plane but the sun is near the horizon. During twilight, particles in the atmosphere will make the distribution of solar radiation more even, allowing a relatively large area of the sky to remain bright, and the ratio decreases again.

3.1.8 Measuring fluence rate and radiance

Spherical sensors, as needed for measuring fluence rate, also called scalar irradiance, are not common. LI-COR (Lincoln, NE, USA) sells a spherical PAR quantum sensor (LI-193) that can be submerged, and Biospherical Instruments Inc. (San Diego, CA, USA) makes both PAR- (e.g. QSL-2100 and QSL-2200, terrestrial; QSP series with models that can be submerged to thousands of metres) and also narrowband spherical sensors for measuring UV- and visible radiation fluence rate. Biospherical Instruments Inc. also makes radiance sensors with input optics that have a very narrow acceptance angle. Figure 3.3 shows the different entrance optics available for one series of sensors from Biospherical Instruments. Most broadband UV sensors and entrance optics for spectroradiometers follow a cosine response. Björn (1995) describes a method for estimating fluence rate from three or six irradiance measurements at a series of specific angles.

3.1.9 Sensor output

Radiometer sensors can have either analog or digital outputs. Some sensors with an analogue output have an amplifier next to the detector, others do not. Sensors with an analog output are usually connected to a voltmeter or a datalogger. Some digital sensors are really complete radiometers with a digital output. Examples of radiometers with digital outputs, are shown in Figure 3.4. RS-232 and USB are digital interfaces frequently used to attach these sensors to personal computers[1].

3.1.10 Calibration

To calibrate radiation sensors the relationship must be determined between the electrical signal produced by a sensor and the amount of radiation impinging on that sensor. The physical value of irradiance, E, of the incoming radiation in energy units, is obtained by "comparison" of a measurement X with that of a calibrated radiation source X_{lamp},

$$E = E_{\text{lamp}} \cdot \frac{X \cdot a}{X_{\text{lamp}}} \qquad (3.3)$$

where E_{lamp} is the calibration file of the lamp provided by the calibration survey. The factor a accounts for any difference in the lamp to sensor distance between the survey's calibration and your own measurement of the lamp. In most modern instruments this "comparison" is implicitly done by the software in the instrument itself, or the computer it is attached to, by multiplying the electrical signal from the detector with a calibration constant. If this is not the case, it is necessary to correct all raw measurements, here X and X_{lamp}, before making any further calculations, for example, by subtracting the measurement of dark current (the sensor reading in the dark).

Although photons can be counted using a photomultiplier and appropriate electronics, this approach does not provide an absolute measurement. Some photons are always missed and false counts are included due to thermal excitation. Therefore, absolute calibration of radiation meters can only be provided in energetic terms. For this purpose so-called "blackbody radiators" of known temperature are used, since they depend only on temperature for the total radiation as well as its spectral distribution. Blackbody radiators used for the calibration of lamps, are then available for purchase by scientists for use as secondary standards.

In the case of broadband sensors used to measure biologically-effective irradiances or selected bands of the spectrum, the calibration is usually carried out by comparison to readings from a calibrated spectroradiometer under a radiation source with a spectrum as similar as

---

[1]Many current computers, especially laptops, do not have an RS-232 interface but USB to RS-232 adaptors are readily available.

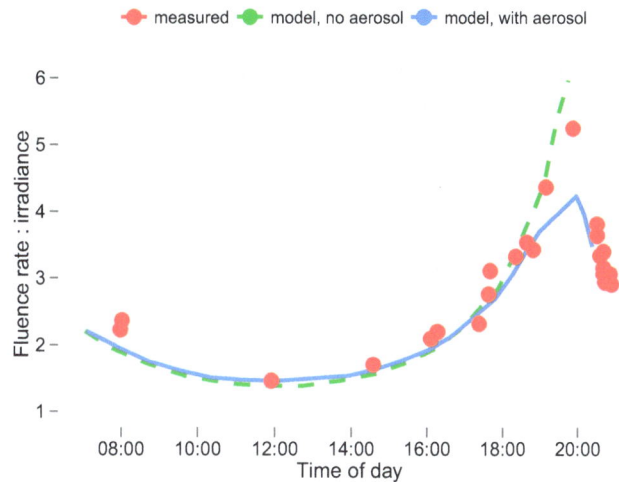

Figure 3.2: Ratio of fluence rate to irradiance throughout a summer day in southern Sweden under clear-sky conditions. The graph presents photosynthetically active radiation (PAR, 400–700 nm). For UV-B the diffuse component is proportionally larger than for PAR due to scattering in the atmosphere (see section 1.4 on page 8 and Figure 1.6 on page 12) and consequently the ratio at midday is also larger and the variation with time smaller. Under overcast sky conditions diffuse radiation predominates and variation through the day is smaller than under clear sky conditions. Redrawn from Björn and Vogelmann (1996).

Figure 3.3: Different entrance optics used in USB radiometers of the 'AMOUR' series. From left to right: radiance, fluence rate (or scalar irradiance), irradiance and SMA connector. The SMA connector can be used to attach an optical fibre. Photograph © by Biospherical Instruments Inc.

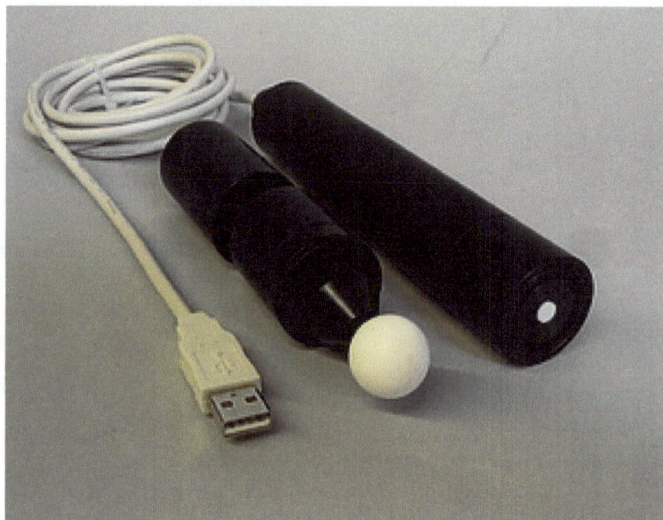

Figure 3.4: 'AMOUR' USB radiometers. Left: With a spherical diffuser for measuring fluence rate; right: with a planar cosine diffuser for measuring irradiance. Photograph © by Biospherical Instruments Inc.

possible to the one which will be measured with the broadband sensor (e.g. sunlight). To calibrate spectroradiometers, in addition to a spectral-irradiance calibration against one or more lamps (with a continuous spectrum, such as an incandescent lamp), it is necessary to do a wavelength calibration against the sun or a lamp (with a an emission spectrum with discrete, narrow and stable peaks, such as a low pressure mercury lamp).

3.1.11 Further reading

Björn and Vogelmann (1996) treat the same subjects as this section, but in more depth. Also the book edited by Björn (2007) is a good source of basic information about radiation and photobiology.

3.2 Actinometry

Actinometers are chemical systems for the measurement of light and ultraviolet radiation. They do not need to be calibrated by the user, and thus do not require the purchase of an expensive standard lamp with an expensive power supply. Standardization has usually been taken care of by those who have designed the actinometer. Another advantage is that their geometry can more easily be adjusted to the measurement problem. The shape of a liquid actinometer can be made to correspond to the overall shape of the irradiated object under study.

In many cases, it is of interest to study a suspension or solution that can be put in an ordinary cuvette for spectrophotometry or fluorimetry, and the actinometer solution can be put into a similar cuvette. A large number

of actinometers have been devised. Kuhn et al. (1989) lists, briefly describes and gives references for 67 different systems of which they recommend five. In general, actinometers are sensitive to short-wave radiation (<500 nm) and insensitive to long-wave radiation (>500 nm). Insensitivity to long-wave radiation can be both a drawback and an advantage, but by choosing the best actinometer for a particular purpose we can avoid their disadvantages. One advantage of using an actinometer insensitive to long wave radiation is that we can use it for UV work under illumination visible to the human eye, without disturbing the measurement. Here we shall concentrate on the most popular actinometer for ultraviolet radiation—the potassium ferrioxalate or potassium iron(III) oxalate actinometer. In addition to using it directly in some experiments, we can use it for checking the calibration of other instruments, such as spectroradiometers.

The description below is sufficient for a researcher starting to work in the field. For more detailed information one should consult Goldstein and Rabani (2008); Hatchard and Parker (1956); Lee and Seliger (1964); Parker (1953). Complete recipes have also been published, e.g., Jagger (1967); Seliger and McElroy (1965). In the ferrioxalate actinometer the following photochemical reaction is exploited:

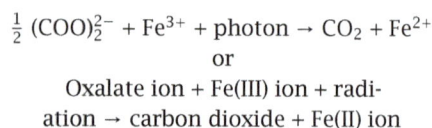

$$\tfrac{1}{2}\,(COO)_2^{2-} + Fe^{3+} + \text{photon} \rightarrow CO_2 + Fe^{2+}$$
$$\text{or}$$
Oxalate ion + Fe(III) ion + radiation → carbon dioxide + Fe(II) ion

The quantum yield for this reaction (i.e., the number

of iron ions reduced per photon absorbed) is slightly wavelength dependent but close to 1.26 in the spectral region, 250–500 nm, where the ferrioxalate actinometer is used. Usually a 1-cm layer of 0.006 M ferrioxalate solution is used. Quantum yield and the fraction of radiation (perpendicular to the 1 cm layer) absorbed are shown in Table 3.1.

As seen from Table 3.1 the sensitivity of this actinometer (column to the right) is constant throughout most of the UV range, which makes it very convenient for our work.

The amount of Fe(II) formed can be measured spectrophotometrically after the addition of phenanthroline, which gives a strongly absorbing yellow complex with Fe(II) ions.

The ferrioxalate (actually potassium ferrioxalate) for the actinometer is prepared by mixing 3 volumes of 1.5 M potassium oxalate (COOK$_2$) with 1 volume of 1.5 M FeCl$_3$ and stirring vigorously. This step and all the following procedures involving ferrioxalate should be carried out under red light (red fluorescent tubes or LEDs). The precipitated K$_3$Fe(C$_2$O$_4$)$_3 \cdot$3H$_2$O should be dissolved in a minimal amount of hot water and the solution allowed to cool for crystallization (this crystallization should be repeated twice more). Potassium ferrioxalate can also be purchased ready-made, but the price difference encourages self-fabrication. The following is a recipe for the three solutions required to carry out actinometry (see Goldstein and Rabani, 2008, for a different procedure and other quantum yields):

Solution A: Dissolve 2.947 g of the purified and dried K$_3$Fe(III) oxalate in 800 ml distilled water, add 100 ml 0.5 M sulfuric acid, and dilute the solution to 1000 ml. This gives 0.006 M actinometer solution, which is suitable for measurement of ultraviolet radiation.

Solution B: The phenanthroline solution to be used for developing the colour with Fe(II) ions should be 0.1% w/v 1:10 phenanthroline monohydrate in distilled water.

Solution C: Prepare an acetate buffer by mixing 600 ml of 0.5 M sodium acetate with 360 ml of 0.5 M H$_2$SO$_4$.

Solution A is irradiated with the radiation to be measured. The geometries of both the container and of the radiation are important and must be taken into account when evaluating the result. The simplest case is when the radiation is collimated, the container a flat spectrophotometer cuvette, the radiation strikes one face of the cuvette perpendicularly, and no radiation is transmitted. Even in this case one has to distinguish between whether the cuvette or the beam has the greater cross section, and correct for reflection in the cuvette surfaces. The irradiation time should be adjusted so that no more than 20% of the iron is reduced (this corresponds to an absorbance of about 0.66). In the following we shall assume that we

use an ordinary fused-silica or quartz spectrophotometer cuvette with 10 mm inner thickness and containing 3 ml actinometer solution.

After the irradiation and mixing of the actinometer solution, 2 ml of the irradiated solution is mixed with 2 ml of solution B and 1 ml of solution C, and then diluted to 20 ml with distilled water. After 30 minutes the absorbance at 510 nm is measured against a blank made up in the same way with unirradiated solution A. An absorbance of 0.5 corresponds to 0.905 µmol Fe^{2+}. It is a good idea to check this relationship with known amounts of FeSO$_4$ if you have not previously checked your spectrophotometer for accuracy and linearity. You should not use any absorbance above 0.65.

Example of calculation: 3 ml of 0.006 M actinometer solution are irradiated by parallel rays of 300 nm UV-B impinging at right angles to one surface (and not able to enter any other surface). The radiation cross section intercepted by the solution is 2 cm^2. Five minutes of irradiation produces an absorbance of 0.6. This corresponds to $0.6 \cdot 0.905/0.5$ µmol $= 1.086$ µmol Fe^{2+}, but since we have taken 2 out of the 3 ml actinometer solution for analysis, multiply by 3/2 to get the total amount of Fe^{2+} formed. Throughout the UV-B region the quantum yield is 1.26, so this corresponds to absorption of $3 \cdot 1.086/2/1.26$ µmol photons. Reflection from the surface is estimated to be 7% (by application of Snell's law, or law of refraction, giving the angle of refraction for an angle of incidence at the boundary of two media like water and glass). None of the radiation penetrates the solution to the rear surface, since the solution thickness is 1 cm. Therefore the incident radiation is $3 \cdot 1.086/(1.26 \cdot 0.93 \cdot 2)$ µmol $= 1.390$ µmol radiation incident on 2 cm^2 in 5 minutes, and the photon irradiance (quantum flux density, in this case equal to the photon fluence rate, since the rays are parallel and at right angles to the surface) is $1.390/(2 \cdot 5)$ µmol/cm^2/min $= 0.1390$ µmol/cm^2/min $= 10000 \cdot 0.1390/60$ µmol/m^2/s $= 23.2$ µmol/m^2/s.

Kirk and Namasivayam (1983) point out errors that might arise if a more concentrated actinometer solution is used, in order to absorb more light at long wavelengths, and how these errors can be minimized. If an actinometer much more concentrated than 0.006 M is used, the quantum yield is lower, and we do not recommend this for UV research. Goldstein and Rabani (2008), using 0.06 M actinometer solution, find almost the same quantum yield (1.24) from 250 to 365 nm, but much higher (1.47) from 205 to 240 nm; the latter in marked contrast to the values of Fernández et al. (1979) in the table above, so measurements below 250 nm should be regarded with caution. Bowman and Demas (1976) warn against exposure of the phenanthroline solution to UV,

Table 3.1: Quantum yield and the fraction of radiation (perpendicular to 1-cm layer) absorbed by a 0.006 M ferrioxalate solution

Wavelength, nm	Quantum yield	Fraction absorbed	Quant. yield · fract. abs.
222	0.50[†]		
230	0.67[†]		
238–240	0.68[†]		
248	1.35[†]		
253.7	1.26±0.03	1	1.260
300.0	1.26	1	1.260
300.0	1.26	1	1.260
313.3	1.26	1	1.260
334.1	1.26	1	1.260
363.8	1.270[‡]		
363.8	1.294[‡]		
365.6	1.26	1	1.260
404.7	1.16	0.92	1.067
406.7	1.188[‡]		
435.0	1.11	0.49	0.544
509.0	0.85	0.02	0.017

[†] Values from Fernández et al. (1979); [‡] Values from Demas et al. (1981); other values based on Hatchard and Parker (1956) and Lee and Seliger (1964).

and even against the fluorescent room lighting.

Chemical or biological systems (mostly in a solid state) for recording solar radiation and particularly UV radiation, are widely employed for estimating the exposure of people, leaves in a plant canopy, and other objects which for various reasons are not easily amenable to measurements with electronic devices. These chemical devices are generally referred to as dosimeters rather than actinometers, even if there is no defined delimitation between these categories. As the construction, calibration, and use of chemical and other dosimeters have been the subject of frequent reviewing (Bérces et al., 1999; Horneck et al., 1996; Marijnissen and Star, 1987), we shall not dwell on them here, only stress that their radiation sensitive components can be either chemical substances (natural such as DNA or provitamin D, or artificial) or living cells (e.g., various spores and bacteria).

3.3 Dosimeters

Broadband dosimeters have been developed to quantify exposure to UV radiation based either upon the photochemical degradation of chemical compounds and plastic films or using biological techniques involving damage to DNA. The range of experimental methods has been described by Dunne (1999) and Parisi, Turnbull et al. (2010). The most practical and effective dosimeters for plant studies include the use of plastic films of polysulphone (PS) and poly 2,6-dimethyl-1,4-phenylene oxide (polyphenylene oxide or PPO) (see Geiss, 2003; Parisi, Schouten et al., 2010) and the determination of spore viability after UV exposure (see e.g. Furusawa et al., 1998;

Quintern et al., 1997, 1992, 1994).

A commercially available UV-dosimetry system 'Viospor' (Biosense, Germany) uses the DNA molecules of microbial spores immobilised in a film mounted in a protective casing with a cosine corrected filter system to provide a measurement of biologically-weighted UV exposure (Figure 3.5). Viospor sensors are available as two types: Viospor blue-line types I-IV which provide estimates of the CIE erythemal exposure (as MED, J m-2, and SED) at a range of exposure levels from seconds to several days, and Viospor red-line which use the DNA damage action spectrum (Setlow, 1974) to estimate the DNA damaging capacity of UV-B and UV-C radiation and the efficiency of UV-C germicidal lamps. After exposure, films are incubated in bacterial growth medium to stimulate spore germination and the production of proteins that are stained for densitometric quantification. Exposed dosimeters are returned to the supplier for analysis (**BioSense**, Dr. Hans Holtschmidt, Laboratory for Biosensory Systems, Postfach 5161, D-53318 Bornheim, Germany. phone: +49-228-653809, fax: +49-228-653809, mailto:mail@biosense.de, internet: http://www.biosense.de).

Small UV dosimeters have also been constructed from 30–45 μm film of the thermoplastic polysulphone and can be used to determine exposure by measuring the increase in absorbance at 330 nm (Geiss, 2003; Parisi, Turnbull et al., 2010) ideally using an integrating sphere to minimise the effects of scattering (Figure 3.6). Dosimeters can be calibrated in sunlight by comparison with erythemally-calibrated broadband radiometers or against lamp sources using a double monochromator spectroradiometer. Ideally, the calibration should be determined

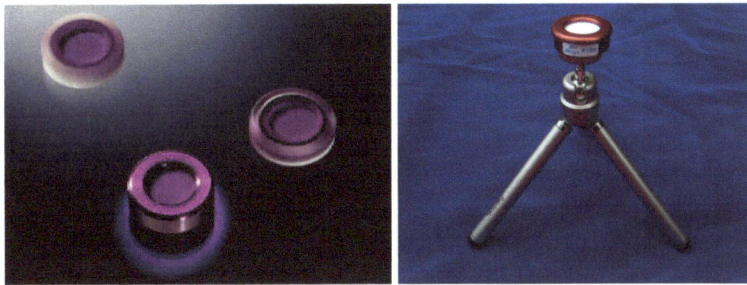

Figure 3.5: Examples of *Viospor* UV dosimeters (Source: Biosense, `http://www.biosense.de`).

under field conditions appropriate for plant studies and if calibrated outside in sunlight the calibration is only accurate under the prevailing atmospheric ozone column as this modifies the UV spectrum. An erythemal dose can be calculated from 40 µm polysulphone film using a relationship of the form (see Geiss, 2003):

$$\text{Radiation amount } (\mathrm{J\,m^{-2}}) = 8025\,(\Delta A_{330})^2 + 1980.8\,\Delta A_{330} \tag{3.4}$$

where ΔA_{330} is the absorbance at 330 nm before exposure (which should be between 0.105 and 0.133) minus the absorbance at 330 nm after exposure plus a further 24 h in the dark. The film may also be calibrated against other action spectra.

Accuracy of polysulphone dosimetry has been reported to be $\pm 10\%$ if ΔA_{330} is <0.3 but decreases to $\pm 30\%$ for ΔA_{330} up to 0.4 (Diffey, 1987). Greater variability in ΔA_{330} occurs with increasing duration of exposure and dosimeters saturate at sub-tropical sites in less than one day. However, they have also been modified with a filter to provide an extended dynamic range of exposure (over 3 to 6 days) without the need to replace the dosimeter due to saturation (Parisi and Kimlin, 2004). Polysulphone dosimeters have been combined with a PAR dosimeter to investigate the visible and UV radiation environment of plants (Parisi et al., 2003, 1998) and miniature versions have also been developed: 1.5 cm × 1.0 cm with an exposure of a 6 mm disc of polysulphone (Parisi, Turnbull et al., 2010).

For longer exposure periods, dosimeters using an alternative plastic film, PPO, have been found more suitable as they saturate at sub-tropical locations after 5–10 days. The change in absorbance of PPO is quantified at 320 nm and it has been successfully calibrated to erythemal exposures (Lester et al., 2003) and by using a mylar (polyester) filter for estimation of UV-A exposures (Turnbull and Schouten, 2008). Both PS and PPO dosimeters have been investigated for underwater use where PPO has been considered viable when calibrated under water (but not using a calibration in air) and under the relevant ozone

column conditions of the study (Schouten et al., 2007, 2008). The duration of use of PPO dosimeters in air at sub-tropical locations has been extended from 5 to 10 days by the use of neutral density filters (Parisi, Schouten et al., 2010; Schouten et al., 2010).

The use of properly calibrated UV dosimeters can be particularly valuable in plant studies when long-term use of spectroradiometers and broadband radiometers is restricted by availability of electrical supplies or by physical constraints.

3.4 Thermopiles

Most thermopiles have a flat response to (energy) irradiance across a wide range wavelengths. They are arrays of thermocouples formed between two different metal alloys. In a thermopile some couples are painted white and some black (or some other arrangement is used to generate a temperature difference dependent on absorbed energy), and the difference in temperature induced by the absorption of radiation by the black regions generates an electrical signal. A single thermocouple produces a very small signal, but connecting them in series generates a large signal that is easier to measure. Thermopiles can be either small for use in the laboratory or larger, and protected by a quartz dome for use in the field. Thermopile pyranometers are used to measure solar radiation in the range 285 to 2800 nm. Examples of such instruments are the pyranometers in the CMP series from Kipp & Zonen. Thermopile pyranometers are standard instruments in weather stations. Thermopiles can be also used to measure the (energy) irradiance of monochromatic radiation, including UV radiation, if the dome or 'window' is made from an UV transparent material.

3.5 Broadband instruments

Broadband radiometers are instruments used to measure irradiance over a broad region of the solar (or lamp) spectrum, weighted with an instrument spectral response.

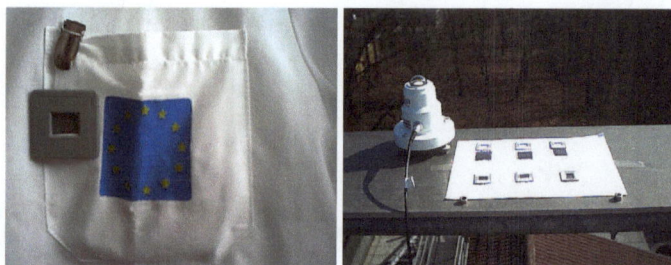

Figure 3.6: Examples of polysulphone dosimeters and their calibration with a broadband radiometer (source: Geiss, 2003).

A broadband radiometer consists of a sensor that is designed to measure the solar radiation flux density $(W\,m^{-2})$ from a field of view of 2π steradians (180°, or π radians, when projected on a perpendicular plane). The plane of the sensor is usually positioned horizontally, but it can also be located, for example, parallel to the surface of a leaf.

UV broadband radiometers integrate over either the UV-A or UV-B band or both, which encompasses the entire UV region of daylight. The names broadband and narrowband refer to the width of the 'window' or range of wavelengths to which a sensor responds. The term full-width half-maximum (FWHM) is used to measure this, it means the width of the peak in units of wavelength, measured at half the maximum height of the peak along the y-axis (with the output of the sensor on the y-axis plotted on a linear scale). A narrow band-pass can have a 10 nm or 20 nm FWHM while a wide one can have an 80 nm FWHM for instruments measuring a combination of UV-A and UV-B radiation.

Their low cost, fast response (typically milliseconds to seconds), stability and low maintenance requirements make them suitable for continuous monitoring applications. The most common spectral response is one that follows the erythemal action spectrum defined by the Commission Internationale de l'Eclairage, or CIE, (McKinlay and Diffey, 1987; A. R. Webb et al., 2011), which describes the response of the human skin to UV radiation (Figure 1.19, on page 25).

However, erythemally effective UV can be derived from most UV measuring instruments if the radiation spectrum is known and fairly stable in time such as when measuring sunlight. Hence, in meteorology the UV index is taken as a common factor that should be obtained from the data at every UV measuring site. Vice versa, using a correction factor to the instrument's output, actual erythemal irradiance in effective $W\,m^{-2}$ can be calculated. Data from UV broadband instruments are part of a worldwide UV database that is located in the World Ozone and Ultraviolet Radiation Data Centre (WOUDC) as a part the Global Atmosphere Watch (GAW) programme of the World Meteorological Organization (WMO).

As the spectral response of broadband radiometers only approximately follow the BSWF needed for the desired response, their use for measuring different UV radiation sources requires source-specific calibration. The readings of broadband radiometers calibrated for sunlight should not be used to assess doses from lamps without using correction factors obtained by calibration. For measuring lamps and LEDs the use of a spectroradiometer is strongly recommended.

3.5.1 Principle of operation

The basic design of broadband instruments has not changed significantly since the introduction of the first erythemally weighted solar UV radiometer, the Robertson-Berger meter (Berger, 1976; Robertson, 1972). When direct and scattered solar radiation is transmitted through the UV transmitting quartz dome, the most common way to obtain an erythemal weighting is to filter out nearly all visible light using UV-transmitting black-glass blocking filters. The remaining radiation then strikes a UV sensitive fluorescent phosphor to convert UV-B light to visible light, i.e. green light emitted by the phosphor. This light is filtered again using coloured glass to remove any non-green visible light before impinging on a gallium arsenic or a gallium arsenic phosphorus photodiode used as detector.

A thermally stable amplifier converts the diode's output current to a voltage. It drives a line amplifier that provides a low impedance 0 to +4 V DC output signal. Phosphor efficiency decreases by approximately $0.5\%\,K^{-1}$ and its wavelength response curve is shifted by approximately 1 nm towards the red every 10 K. This latter effect is particularly important because of the slope of the solar radiation curve at these wavelengths. The glass filters, phosphor and photodiode are held at 25 to 50°C, depending on the manufacturer, to ensure that the output signal is not sensitive to changes in ambient temperature. Temperature stabilization is usually achieved by an internal thermistor that permits independent monitoring of the

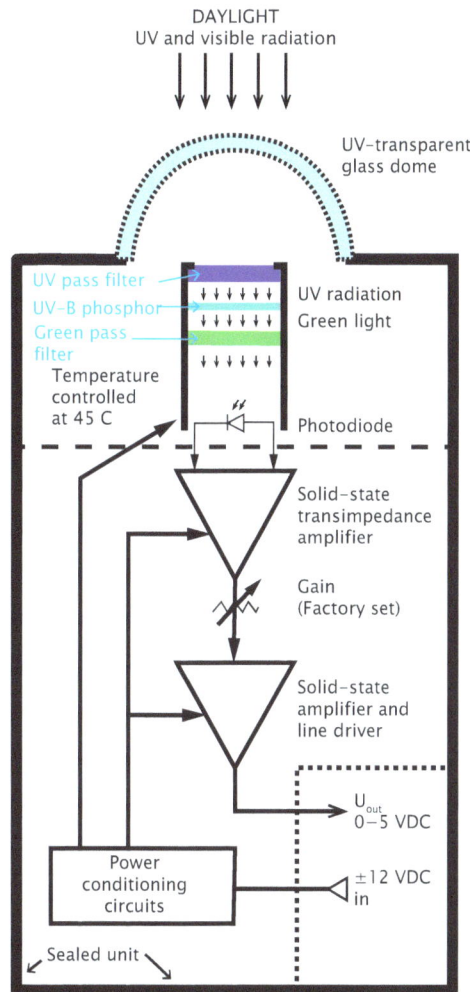

Figure 3.7: Schematic drawing of the Yankee UV pyranometer. Redrawn from diagram © Yankee Environmental Systems.

sensor's temperature (Figure 3.7). The analogue electrical signal produced by the broadband UV instrument is converted into digital format for electronic logging. The sampling frequency is usually between once per second (1 Hz) and once per minute. If the complete data set is not stored then data are saved as averages over periods ranging from 10 minutes to not more than one hour. Sometimes, the variation around the mean is also recorded for each averaging period. This indicates the constancy of the conditions during the averaging period (e.g. sun screening by rapidly changing cloud cover: broken clouds or clear sky or constant cloudiness). The raw signal must be converted into units of erythemal irradiance ($W m^{-2}$) using a calibration factor, plus several corrections. These corrections require additional data: solar zenith angle (θ)

and ozone column depth at the time of measurement.

Erythemal effective irradiance (E_{CIE}) is calculated (A. Webb et al., 2006):

$$E_{CIE} = (U - U_d) \cdot k \cdot f(\theta, \omega) \cdot f(T) \cdot \varphi \qquad (3.5)$$

Where: U is the measured electrical signal from the radiometer, U_d is the electrical offset for dark conditions, k is the calibration coefficient, a constant value determined for specific conditions, e.g. at θ of 40° and a total ozone column of 300 DU[2]. $f(\theta, \omega)$ is a function of the solar zenith angle (θ) and the total column of ozone (ω), i.e. the function can be expressed as a calibration matrix (or look up table) and is derived as part of the calibration procedure. It is normalized at a solar zenith angle of 40° and a total ozone column of 300 DU. For solar zenith

[2]Dobson unit: the depth of a column of pure ozone, at ground level and a temperature of 0°C, equivalent to the ozone content in the atmosphere; 300 DU = 3 mm of pure O_3.

angles less than 40°, $f(\theta, \omega)$ is often nearly unity. $f(T)$ is the temperature correction function. It is recommended that the instrument is temperature stabilized. If this is not applicable then a correction should be applied, which is complex and not always successful. φ is the cosine correction function (if necessary, otherwise =1).

The quality of the broadband instrument depends on the quality of the protective quartz dome, the cosine response, the temperature stability, and the ability of the manufacturer to match the erythemal curve with a combination of glass and diode characteristics. Instrument temperature stability is crucial, with respect to both the electronics and the response of the phosphor to the incident UV radiation.

More recently, broadband instruments are using thin film metal interference filter technology and specially developed silicon photodiodes to measure UV erythemal irradiance. This overcomes many problems connected with the phosphor technology, but on the other hand they have difficulties related to very low photodiode signal levels and filter stability. Silicon carbide (SiC) photodiodes have good sensitivity to UV radiation and are intrinsically blind to visible radiation.

Other broadband instruments use one of these measurement technologies to measure other regions of the UV spectrum by using either a combination of glass filters or interference filters. Some manufacturers of these instruments provide simple algorithms to approximate erythemal dosage from the unweighted measurements (WMO, 2008).

The maintenance of broadband instruments consists of ensuring that the domes are cleaned, the instrument is level, the desiccant (if provided) is active, and the heating/cooling system is working correctly, if so equipped.

3.5.2 Some commonly used terrestrial instruments

The most common outdoor broadband radiometers are: SL 501 from Solar Light, Inc. (Glenside, PA, USA), YES UVB-1 from Yankee Environmental Systems, Inc. (Turners Falls, MA, USA), UVS-E-T (erythemal), UVS-A-T (UV-A) and UVS-B-T (UV-B) from Kipp & Zonen (Delft, The Netherlands) (Figure 3.8). The principle of operation of these meters is basically the same as described in section 3.5.1. Unlike the other meters, the Scintec UV-S-290-T uses a Teflon diffuser under the quartz dome in front of the filters. These three instruments are temperature stabilized by means of heating elements and temperature sensors. Solar light also sells non-stabilized instruments like the PMA2101 (digital) and PMA1101 (analog), which contain a temperature sensor whose output can be used to cor-

rect in silico for the temperature dependency of the UV-B readings.

Broadband sensors based on special silicon photodiodes are also available, which, are more stable with respect to variation in temperature than those based on phosphors. Delta-Ohm (Padova, Italy), Delta-T Devices (Cambridge, UK) and Sky instruments (Llandrindod Wells, UK) make sensors based on this principle. Examples are the SKU 420, SKU 430, and SKU 440 (UV-A, UV-B, and erythemal, respectively) from Sky Instruments, LP UVA 01/03 and LP UVB 01/03 from Delta-Ohm, and UV3pB-05 and UV3pA-05 from Delta-T devices (Figure 3.9). All of them are UV radiometers with no temperature stabilization or in-built temperature sensors. International Light Technologies (Peabody, USA) makes a wide array of meters and sensors for measuring UV radiation. Vital Technologies used to make good UV sensors which were popular some years ago, but the company is no longer in business. Most of these sensors have built-in preamplifiers.

3.5.3 Spectral and angular (cosine) response

Radiation incident on a flat horizontal surface originating from a point source with a defined zenith position will have an intensity value proportional to the cosine of the zenith angle of incidence. This is called the 'cosine law' or 'cosine-response' (see section 3.1.2 on page 71). Ideally, a pyranometer has a directional response, which is the same as the cosine-law. Nevertheless, directional response in a pyranometer is influenced by the quality, dimensions and construction of the (quartz) dome and/or Teflon diffuser. Pyranometer cosine-response is defined in their manufacturers specification as deviation from the ideal cosine-response using the incidence angle up to 80° with respect to $1000\,\mathrm{W\,m^{-2}}$ irradiance at normal incidence (0°). Most sensors deviate considerably from ideal cosine response at angles between 80° and 90°.

The erythemal response of human skin to UV radiation varies with the individual, but for the global evaluation of UV-related health effects to succeed, broadband measurements have to be standardized, which means that the radiometric characteristics of all meters should be identical. The spectral response of every meter should follow exactly the same reference action spectrum and the angular response should not deviate from the cosine response (Leszczynski, 2002). Hence, the spectral response of an ideal erythemally weighted radiometer should follow the CIE curve, and the angular response should follow the cosine function. Unfortunately, the angular and spectral response of real erythema meters are far from ideal; moreover, the characteristics vary from one meter unit to another, even within the same meter type (Leszczynski, 2002).

Figure 3.8: UV-B radiometers (phosphor based, temperature stabilized), from left to right: Solar Light SL 501, Kipp & Zonen UVS-B-T, YES UVB-1. © Kipp & Zonen B.V. (center), and © Yankee Environmental Systems Inc. (right)

Figure 3.9: UV-B radiometers (photodiode based, non temperature stabilized), from left to right: Delta-T UV3p sensors and Skye SKU 430 UV-B sensor. © Delta-T (left) & © Skye Instruments (right).

Broadband radiometers that do not follow the CIE erythemal action spectrum as BSWF are also used. For some radiometers the spectral response follows a bell-shaped curve centred on the UV-A or UV-B bands (see Figure 3.10). Radiometers following the GEN or other spectra commonly used as BSWFs in research with plants, are very seldom used, and they are currently not available commercially. However, almost any UV-B radiometer can be calibrated to measure according to these BSWFs, but such calibrations are valid only when the calibration light source exactly matches the spectrum of the measured light source.

3.5.4 Calibration and intercomparison

Each broadband UV instrument used to measure solar or lamp radiation should be characterized for its spectral and angular response, and its sensitivity to temperature (and if possible humidity). These characteristics should be checked at regular intervals to determine their stability. Also, correction is necessary for each instrument, as no instrument has a spectral sensitivity identical to the erythemal action spectrum.

To calibrate a broadband instrument for solar radiation, the basic procedure is to simultaneously measure the spectral irradiance of the sun with a calibrated spectroradiometer and the broadband meter under cloudless sky conditions. The measured spectrum is weighted with the desired spectral sensitivity[3] of the broadband meter and integrated over all wavelengths relevant for the broadband meter. The result is given in the units [detector-weighted $W\,m^{-2}$], relative to a defined wavelength, usually the maximum of the erythemal action spectrum (CIE) at 298 nm or the maximum of the spectral sensitivity of the broadband meter. For different atmospheric conditions, such as different solar elevation or different thickness of the ozone column, the relationship between the detector-weighted spectral integral, measured with a spectroradiometer, and the output of the broadband detector, after cosine correction, should be constant within the uncertainty estimate; if this is not the case, the mismatch indicates that the spectral sensitivity of the broadband meter deviates from the that used to calculate effective irradiances from the spectral irradiance data (e.g.

---

[3]The desired or target BSWF used for calculating the effective dose will differ to some extent from the real spectral response of the sensor. This means that the calibration will depend on the spectrum of the radiation source.

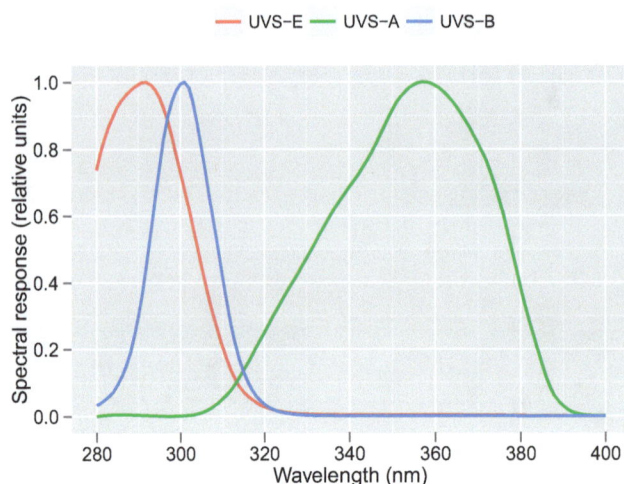

Figure 3.10: Spectral response of three different UV radiometers from Kipp & Zonen. Spectral response is the electrical signal output by the sensor for radiation of different wavelengths, but of the same spectral (energy) irradiance. Data source: manufacturer's brochure.

that defined by the CIE standard for erythema), or that the spectroradiometric measurements were incorrectly done (Seckmeyer et al., 2005).

When measuring the output from UV lamps using broadband sensors calibrated under sunlight, large errors are incurred. In the example shown in Table 3.2, we use Kipp sensors because this manufacturer has published instrument spectral response data plotted on a logarithmic scale. Errors of a similar magnitude should be expected for equivalent sensors from other manufacturers. When measuring the irradiance from lamps under a background of sunlight as in some modulated systems, or when the spectrum changes as occurs when profiling the UV radiation change with depth in water bodies, other types of instruments like multiband sensors or spectroradiometers are preferable.

3.5.5 UV radiation monitoring in growth chambers, greenhouses and phytotrons

Erythemal broadband instruments are widely used to monitor UV radiation levels in growth chambers. Great care must be taken when using such instruments for this purpose since plant action spectra generally deviate from the CIE function. In addition, the reflectance of walls and other surfaces of growth chambers may affect the readings if the cosine response is not good. Great care is needed when artificial light sources are used, because their spectra differ greatly from the solar spectrum for which broadband instruments are normally calibrated.

Correction factors for the solar zenith angle and ozone dependence of the calibration factors are based on unfiltered solar spectra, so cannot be applied to measurements performed in such chambers, hence special treatment of data may be necessary. In most cases if absolute readings are needed, the broadband sensor should be calibrated against a double-monochromator spectroradiometer, for each different light source to be measured. Failing to do so can cause huge errors in the measurements of doses as shown in Table 3.2.

3.6 Multi-channel filter instruments

Multichannel instruments are radiometers that measure a series of fixed, usually narrow, wavelength bands of radiation. They are more rugged and cheaper than high quality spectroradiometers and easier to deploy. Each channel has its own detector (e.g. silicon photodiode) and filter (e.g. interference filters). Usually there is a single diffuser acting as common light collector for all channels. One example of a multichannel instrument is the GUV-2511 from Biospherical Instruments Inc (San Diego, USA) designed to measure cosine-corrected downwelling irradiance at 305, 313, 320, 340, 380, and 395 nm, as well as PAR (400–700 nm). When measuring daylight this allows to monitor UV radiation in key UV wavebands for biological exposure studies. These wavelengths also allow the extraction of cloud optical thickness and total column ozone, two critical variables used in modelling the solar spectrum. Multichannel sensors are mainly

Table 3.2: Simulation of measured values for three common types of sensor. We simulate a perfect calibration for sunlight, and express the 'measured' irradiance by the sensors as a fraction of the true irradiance. We do this for some commonly-used lamp types. We compare three types of lamps, UVB-313 and UVA-340 (Q-Panel) and TL12 (Philips) not filtered, and UVB-313 filtered with cellulose di-acetate, in combination with three types of broadband sensors UVS-E (erythemal), UVS-A (unweighted UV-A), and UVS-B (unweighted UVB) from Kipp & Zonen (see Figure 3.10 for the sensor response spectra used). The sunlight spectrum used for calibration is a daily accumulated value simulated for Jokioinen, Finland, for 21 May, expressed on an energy basis (see Kotilainen et al., 2011, for details). All spectra are simulated for clear sky conditions (CMF=1.0).

	day	8:30	11:30	UVA-340	UVB-313	+ acet.	TL12	BSWF or wavelength band
UVS-E	1.00	1.03	0.95	0.77	0.42	0.47	0.37	CIE98, erythemal
UVS-A	1.00	1.00	1.00	1.26	0.93	0.94	0.92	unweighted UV-A energy irradiance
UVS-B	1.00	0.99	0.99	1.02	1.34	1.34	1.30	unweighted UV-B energy irradiance
UVS-A	1.00	1.00	1.00	1.32	1.02	1.03	1.01	unweighted UV-A photon irradiance
UVS-B	1.00	0.99	1.00	1.02	1.37	1.36	1.34	unweighted UV-B photon irradiance

used for long-term monitoring of UV irradiance and its geographic variation. They are also used for ground measurements used to calibrate space-borne instruments carried by satellites.

Recent multichannel instruments from Biospherical Instruments are modular. They are composed of microradiometers of small size, one for each channel, which together with input optics and filters are used to build the multichannel instruments. Figure 3.11

The ELDONET terrestrial dosimeter consists of three broad band sensors, measuring UV-B, UV-A and PAR irradiance (Figure 3.12). It uses an integrating sphere to collect the light, which reaches the detectors after bouncing on the sphere walls. The autonomous version includes a built-in datalogger. It is waterproof, but it is not submersible. The underwater version is described in section 3.8.3 on page 95.

3.7 Spectroradiometers

Spectroradiometers[4] are instruments used to measure spectral irradiance. There are two types of spectroradiometer: 1) scanning spectroradiometers, and 2) array detector spectroradiometers. The former have a single sensor, which is used to sequentially measure the spectral irradiance at each wavelength, while the latter have an array of sensors onto which the refracted spectrum is projected and measured simultaneously at all wavelengths. Sometimes the more general term spectrometer is adopted to indicate that the same instrument can be also used to measure spectral absorbance, -transmittance, or -reflectance, in addition to spectral irradiance. Scanning spectroradiometers scan a range of wavelengths. To do this they require mechanically moving parts inside the optic path and therefore need very stable housing for the instrument. For this reason, scanning spectroradiometers are larger and more difficult to transport than array de-

tector spectroradiometers. They are also less rugged and usually require mains power. However, in scanning spectroradiometers it is possible to use a double monochromator arrangement that makes their optical performance far superior to that of array spectroradiometers which always use a single monochromator.

3.7.1 Scanning spectroradiometers

3.7.1.1 Basic structure and principles of operation

The basic components of a scanning spectroradiometer are the following: a) input optics for collecting radiation from the sky and guiding it further into the spectroradiometer b) a monochromator for resolving the input radiation into separate wavelengths c) a photomultiplier tube (PMT) for detecting the energy possessed by each spectral component in the measured spectrum. In addition, an external computer for communication with the microprocessor of the spectroradiometer and for collection, processing and storage of data is needed.

The input optics typically consists of a flat teflon diffuser covered by a quartz dome. The diffuser collects the incident photons from the overhead hemisphere. The resulting diffuse radiation is guided to the entrance slit of the monochromator, sometimes by means of an optical fibre. The monochromator may be a single or a double monochromator (see section 2.2.4 on page 44). In scanning spectroradiometers, a system based on a step motor drives a mask that allows only photons of a certain wavelength at a time to enter the exit slit of the monochromator. The exit slit serves as the entrance window to the cathode of the PMT. The photon pulses are amplified and transmitted to a photon counter for registration.

Some spectroradiometers are constructed on a solar tracker that follows the position of the sun. This eliminates the effects of potential azimuthal dependencies in the detection of radiation. A measurement head at

---

[4]It is necessary to distinguish between different types of instruments used to measure spectra or spectrometers. A spectrophotometer is an instrument used to measure optical properties of objects while a spectroradiometer is used to measure radiation.

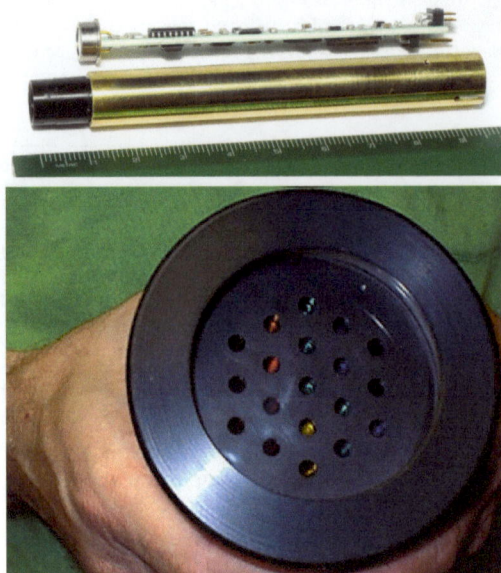

Figure 3.11: Top: microradiometer used in multichannel instruments from Biospherical Instruments; bottom: radiance fore-optics of a PRR-800 radiometer with 19 channels. Photographs © by Biospherical Instruments Inc.

Figure 3.12: ELDONET terrestrial dosimeter. (Photograph by Donat Häder, reproduced with permission.)

the end of an optical fibre may be also installed on a separate sun tracker. The temperature of the instrument is usually either stabilized or kept above a certain temperature limit to ensure proper functioning. The dome of the measurement head may be equipped with a heater and/or air blower to keep the temperature of the teflon within certain limits and to avoid emergence of frost onto the dome.

Other spectroradiometers are less rugged and are intended for laboratory or spot measurements outdoors. They are more portable but more sensitive to temperature extremes and are not water proof. Examples of such instruments are Optronics OL 756 and Macam SR9910 spectroradiometers. Figure 3.13 shows the different parts of the first of these instruments.

The more rugged instruments, usually permanently installed at a fixed location, are commonly used for measuring long (several years long) time series of UV spectral irradiance data. The more portable instruments are used for spot measurements in plant canopies, and under lamps, and or filters. The first type of instrument is most commonly used by meteorologists, while the portable instruments are most useful to biologists.

3.7.1.2 Characteristics

Dark current and **dead time** are characteristics possessed by the PMT. Dark current is a measure of the drift photons going from the cathode to the anode of the PMT without any real incident photons entering the instrument. Dead time is a measure of that PMT which is in a paralysed state after a photon detection event. Stray light is composed of photons echoed from wavelengths other than the nominal wavelength being measured. In commercially available scanning spectroradiometers, these phenomena are usually measured and handled by the measurement software.

The **wavelength alignment** of a spectroradiometer has to be checked regularly. Most instruments taking daily measurements contain an internal mercury lamp aimed at ensuring the stability of the alignment. The wavelength and the position of the micrometer turning the grating of the monochromator are related to each other by a second-order equation using so called dispersion coefficients. The determination of the dispersion coefficients should be part of the annual maintenance of the instruments.

Solar irradiance spectra sometimes exhibit so called **noise spikes**, which mean sudden abnormally high or low intensity readings on a single wavelength. The origin of the spikes is not fully resolved, but straylight is considered a partial explanation. The spikes can be detected and eliminated making use of suitable reference spectra.

Ideally, the angular response of the measurement head follows the shape of a cosine curve. In practice, the response deviates somewhat from this. Typically, the larger the solar zenith angle, the larger the deviation. The **cosine response** of the measurement head should be measured in laboratory and a corresponding correction applied to all measured data.

If the spectroradiometer is not stabilized for temperature, its response usually exhibits **temperature dependence**. This dependence should be determined in the laboratory by measuring a calibration lamp with a spectroradiometer heated/cooled over a range of different wavelengths. The measurements can be used for deriving the temperature correction factors to be applied in the post-processing of the sky measurements.

The **slit function** determines the transmittance of a monochromator as a function wavelength. The ideal shape of the function would be triangular. The full width at half maximum of the slit function is commonly used as a quantity characterizing the slit function. The slit function can be derived by measuring the irradiance emitted by a tunable laser. Removal of the effect of the slit function on the measured spectra should be considered if spectra measured by two or several instruments are to be compared with each other.

The **spectral responsivity** of a spectroradiometer should be based on regular measurements of a certified calibration lamp. If the responsivity seems to have changed, basically two alternative ways to handle the change exist. The change may be introduced in the responsivity of the instrument and the processing of the sky measurements as such. A step-wise change in response is hence introduced in the time series of the measurements. Alternatively, a gradually changing response time-series may be defined using a moving average with a suitable time window. In this way, the change in the response is introduced gradually in the time series of the sky irradiance measurements.

3.7.1.3 Maintenance

The maintenance of a scanning spectroradiometer operating in an outdoor environment involves the following practices: a) general daily maintenance; b) checks on the wavelength setting and stability of irradiance scale; c) calibration of irradiance against primary standards in a dark room.

Daily routine maintenance includes cleaning of the quartz dome and checking on the general functioning as well as the correct levelling of the instrument. The quartz dome should also be cleaned/dried after rain or snow. The operator should be familiar with the control software of the instrument. Additional simple routines based on, for instance, selected reference spectra may be

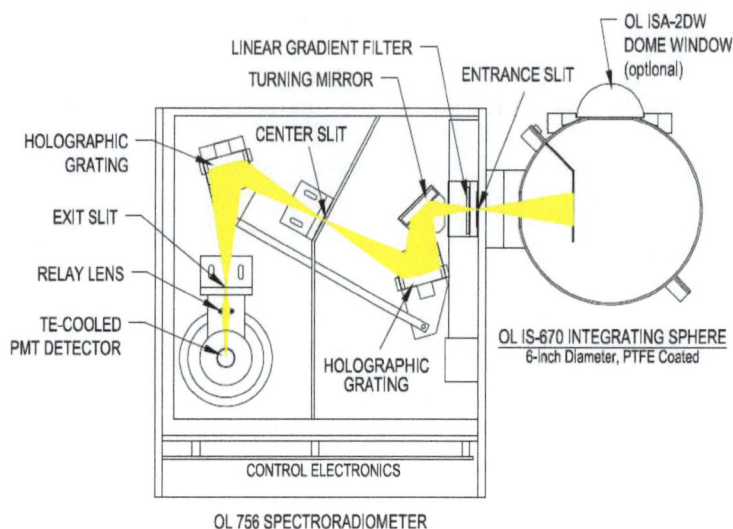

Figure 3.13: Diagram of a double monochromator scanning spectroradiometer, OL 756 from Gooch & Housego, Orlando, Florida, USA. Diagram © Gooch & Housego.

used for instant checking of the measured data. These kinds of routines are invaluable in the prompt detection of occasional malfunctions of the instrument.

An internal Hg lamp is used for checking of the wavelength scale in some spectroradiometers. In these cases, it is convenient to imbed the Hg lamp measurement into the daily measurement schedule. If the instrument lacks an internal lamp, this check has to be done using an external lamp. For checking the stability of the irradiance scale, portable calibration units are available. These enable, for instance, stability checks of the instrument at the measuring site on a weekly basis. It is advisable that the humidity indicators are also checked on a weekly basis.

Irradiance calibration of a spectroradiometer should be performed in a dark room (Figure 3.14). A primary standard lamp with an irradiance certificate provided by a certified laboratory of standards is needed. To extend the lifetime of the primary standard lamp, it is recommended that it is not used as a regular calibration lamp. Instead, the irradiance scale of the primary lamp should be transferred to a secondary standard lamp that is used as a working calibration lamp. Use of several working lamps is recommended to enable recognition of potential drifts in the radiant output of the lamps as they age. Calibration against the primary/secondary standard lamp should be performed at least every two months. The desiccant bags inside the cover of the spectrometer should be taken out and dried at least every two months as well.

Proper levelling of the instrument has to be ensured after having it relocated for outdoor measurements.

On the annual maintenance practices of a spectroradiometer, each manufacturer has its own services and recommendations. Participation in intercomparison campaigns gathering a number of state-of-the-art instruments to conduct measurements on a jointly agreed schedule for a period of time has proven a fruitful way to investigate the long-term stabilities and overall performances of scanning spectroradiometers (Figure 3.15).

3.7.2 Array detector spectroradiometers

In contrast to scanning spectroradiometers, array detector instruments measure spectral irradiance simultaneously at all wavelengths. The detector in this case is a linear array of light sensors, similar in structure to the imaging sensors used in digital cameras, but long and narrow. The number of detector elements ('pixels') along the array varies, 2000 to 3000 pixels being common[5]. The array can be a 'charge coupled device' (CCD), or an array of photodiodes (DAD). The 'image' of the spectrum produced by the monochromator is projected and focused by means of mirrors onto the linear detector array, each detector in the array receiving light of a certain wavelength. In the case of array spectrometers it is not possible to use two monochromators in tandem to reduce stray light. Array spectrometers are small and portable (Figure 3.16).

---

[5] Some instruments with approximately 200 pixel arrays are also available, but these are usually called 'hyperspectral radiometers' instead of spectroradiometers.

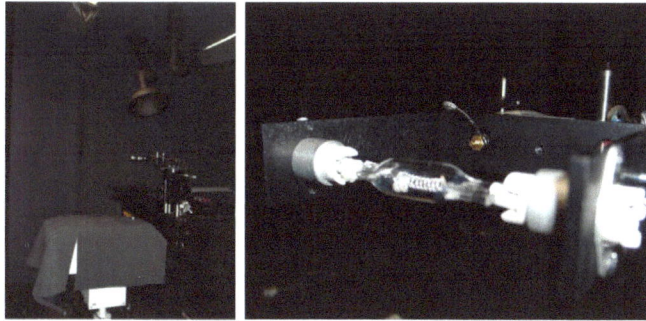

Figure 3.14: Brewer#107 of the Finnish Meteorological Institute in the dark room of Jokioinen Observatory (on the left); a 1000W calibration lamp (on the right). Photograph: Anu Heikkilä.

Figure 3.15: Brewer spectroradiometers participating in a measurement campaign in Huelva, Spain, in October 2005. Photograph: Anu Heikkilä.

Figure 3.16: Left: Maya 2000 PRO spectrometer. Right: The same instrument connected to an optical fibre and a small cosine diffuser. Spectrometer and accessories from Ocean Optics. Photographs: Pedro J. Aphalo.

For measuring energy or photon spectral irradiance a cosine diffuser is used as input optics. This ensures that the angular response follows the cosine law, and so the instrument measures the radiation as received on a flat surface. Other input optics are also available, for example, with a narrow angle of view. However, the quantity measured with them is not irradiance. Cosine diffusers differ widely in how closely they follow the cosine law. Some cheaper models are prone to large errors, especially when radiation impinges at a sharp angle to their surface. This will be further discussed in section 3.7.2.1 on page 91.

The input optics is usually connected to the array spectrometer with an optical fibre. The type of fibre to be used depends on the wavelength range to be measured. If smaller than the entrance of the spectroradiometer, the diameter of the fibre will affect the amount of radiation entering the instrument. The diameter also affects the mechanical properties of the fibre: thin fibres are more flexible and tolerate bending into curves of smaller diameter. Fibres also vary with regards to the type of cladding material used to protect them. Fibres with metal cladding tolerate rougher handling than those with plastic cladding. The most common connector for these fibres and accessories is the SMA 905, originally designed for light fibres used in digital communication systems. For this reason their positioning upon repeated attachment is not exactly the same. Consequently, the recommendation is **not** to detach and reattach the fibre from the spectrometer without recalibrating the system[6].

At the entrance of the spectrometer, just behind the connector to which the fibre is attached, there is a slit (Figure 3.17), which limits the width and height of the incoming light beam. The width is of the order of a few micrometres and the exact value chosen determines, together with the monochromator, the spectral resolution of the spectroradiometer. The narrower the slit, the narrower the beam hitting the monochromator and the better the resolution (the narrower the peaks that can be resolved). In a Czerny-Turner configuration (Figure 3.18), the next component is a collimating mirror which projects the beam onto a monochromator. Gratings are used as monochromators. Gratings have a surface with very closely spaced rulings of a specific profile, and they separate radiation of different wavelengths in a similar way to a prism. One important parameter is the density of rulings which is one of the determinants of spectral resolution and useful wavelength range. The 'image' produced by the grating is focused onto the array detector by another collimating mirror. Some newer models of spectrometer from StellarNet (e.g. BLACK-Comet spec-

trometer) and now also from Ocean Optics (Torus spectrometer) use a concave grating instead of a planar one. Since the grating itself focuses the light onto the array detector, collimating mirrors are not needed. Having fewer optical components, an instrument with better stray light performance is obtained.

The array detector normally has rectangular 'pixels', orientated so that their shorter dimension is on the axis along which the different wavelengths have been separated by the monochromator, and their longer dimension is perpendicular to it. The entrance slit is positioned to have the long dimension coincident with the long dimension of the pixels. In some detectors the long pixels are in reality rows of square pixels with their electrical output combined into a single output signal. The output signal from the pixels is averaged by the detector itself over what is called 'integration time'. The longer the integration used, the lower the irradiance that can be measured. However, the 'dark noise' increases with the integration time. In addition, it is possible to take several scans and average them. A coarse dark noise correction is sometimes done by subtracting the signal from special pixels at the end of the array that are not exposed to radiation. However a dark scan, with the input optics protected from the incoming radiation, should also be measured, and its value, wavelength by wavelength subtracted from the measurements. The dark noise depends on temperature. This has two implications, dark scans should be taken frequently, sometimes before or after each measurement, and the spectrometer should be allowed to warm up for some minutes before starting to take readings. Furthermore, when working outdoors it should be protected from direct sunlight, so as to keep its temperature stable and close to that at which it was calibrated. Some spectrometers have a thermoelement (TE), working according to the Peltier principle, which cools the array detector to a preset temperature and thereby stabilizes it.

Most current array spectrometers, the exception being some models with thermoelectrically cooled detectors, are powered through the USB port of a personal computer. For portable use a laptop is frequently used. Special software, sold by the manufacturer of the spectrometer, is used to control the instrument and acquire and plot the spectra (Figure 3.19). For most instruments there are also drivers and software development kits (SDK) available for developing programs for special applications. When special corrections, for example for stray light, are performed it may be necessary to acquire raw spectral data and apply corrections and calibrations off-line using other software, for example Excel or R.

---

[6]Errors caused by detaching and reattaching a fibre with an SMA connector can be ±5% (Nevas et al., 2012), but the information available is rather limited, so if the fibre is going to be detached, the errors should be characterized for each individual instrument. Errors caused by attaching and reattaching a fibre with an FC connector are negligible (Nevas et al., 2012)

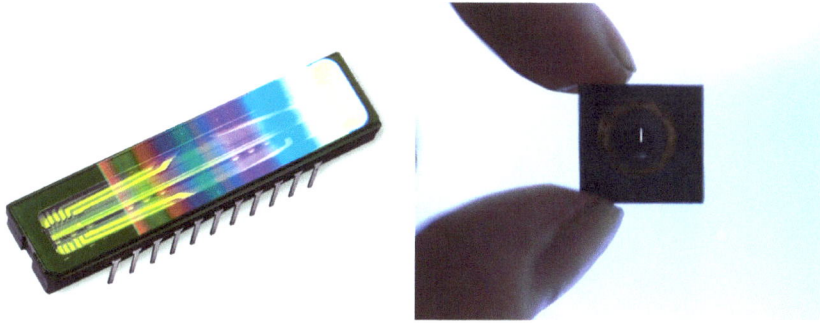

Figure 3.17: Left: A CCD array detector with an order sorting filter applied; right: the slit of an AvaSpec spectrometer. Photographs, Left: © Ocean Optics; right: © Avantes.

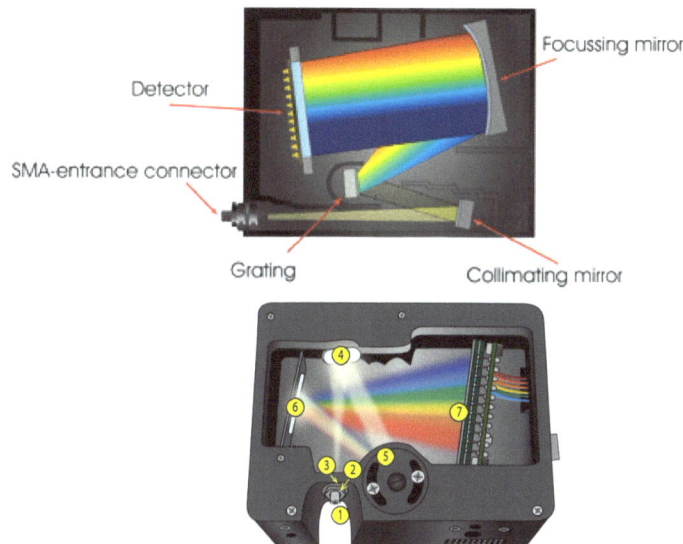

Figure 3.18: Two different layouts for the optical bench of array spectrometers. At the top a symmetrical Czerny-Turner configuration exemplified by the AvaBench used in most Avantes instruments, and at the bottom a Czerny-Turner configuration with a crossed light pass as used in the Ocean Optics USB4000 spectrometer. [1] SMA connector, [2] entrance slit, [3] optional filter, [4] collimating mirror, [5] grating, [6] focusing mirror, [7] detector. Top figure, © Avantes; bottom figure, © Ocean Optics.

3.7.2.1 Measuring errors and limitations in accuracy

Array spectroradiometers have a great advantage when quickly measuring changing radiation as they acquire all wavelengths simultaneously. This ensures that the values of spectral irradiance measured at all wavelengths are consistent. In contrast, under conditions where irradiance varies rapidly with time, the shape of the measured spectrum can get badly distorted when measured with a scanning spectroradiometer. However, array spectrometers have a serious limitation in that they cannot be built with double monochromators. As any spectroradiometer with a single monochromator, they suffer from relatively high values of stray light. Stray light originates from scattered light of incorrect wavelengths falling on a pixel of the array detector. In other words, radiation of one wavelength is detected (and measured) as radiation of a different wavelength. Perfectly scattered radiation would affect all pixels in the same way, but when there are reflections within the optical bench that are not perfectly scattered, some pixels in the array detector are more affected by stray light than others. Stray light is a critical specification when measuring UV-B in sunlight, as UV-B irradiance is very low compared to the irradiance of visible and near infra-red radiation. Consequently, if even a small proportion of visible radiation is scattered and reflected as stray light within the instrument, this stray light can generate a signal on the 'UV-B pixels' of

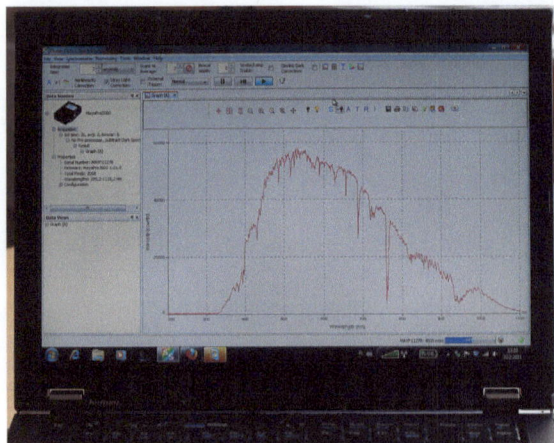

Figure 3.19: Screen shot of the SpectraSuite software from Ocean Optics while measuring a raw spectrum. Photograph: Pedro J. Aphalo.

the array of a magnitude similar to, or larger than, that produced by the UV-B radiation that we are trying to measure. Stray light is such a big problem that without very complicated and special corrections these instruments cannot be used at all to measure UV-B radiation in sunlight. Errors of more than 100%[7] for biologically effective doses can be incurred even with a well calibrated instrument. Failure to take this into account has led to important mistakes, like the erroneous measurement of solar UV-C radiation at ground level by NASA researchers which was published in *Geophysical Research Letters*. This was most likely an artifact due to the limitations of the array spectrometer used. See the paper by D'Antoni et al. (2007) and the refutation by Flint et al. (2008) and the answer by D'Antoni et al. (2008). Equally, the values of the UV-B doses used in many recent biological experiments, as reported in the publications, are suspect, since they have been based on measurements performed with single-monochromator instruments.

Gratings disperse radiation according to what are called 'orders'. For example first order dispersion may be 10 nm/mm, second order dispersion 5 nm/mm, third order dispersion 2.5 nm/mm, and so on. The first order spectrum is what is of interest, and is what we want the array detector to see. However, any given 'pixel' in the array, in addition to radiation corresponding to the first order (e.g. 800 nm), also sees radiation corresponding to higher orders (e.g. 400 nm, 266.6 nm, 200 nm, and so on) if those wavelengths are present in the incoming radiation. The solution to this problem is to use 'order-sorting filters' in the light pass. In array spectrometers order-sorting filters may be directly coated onto the array detector, or attached to it. For example Ocean Optics

spectrometers can be bought with a *variable longpass order-sorting filter* as an option (Figure 3.17).

Another problem with array detector spectrometers is that the radiation may be better focused on some parts of the array than on others, and this causes changes in spectral resolution with changing wavelengths. In addition, the wavelength difference between adjacent pixels is not always the same across the whole spectrum, neither is the step size an integer number. Usually the software supplied with the instrument can generate files with data at integer steps (e.g. 1 nm, or 5 nm) but this is done by interpolating and averaging, rather than changing the measurement itself. In contrast the scanning step of scanning spectroradiometers can be controlled through its software.

The overall accuracy of the measurements is also reliant on the angular response of the entrance optics. For measuring spectral irradiance we generally use a cosine diffusor as entrance optics, although it is also possible to use an integrating sphere. Deviations of cosine diffusers from the theoretical angular response tend to increase at large angles from the vertical. If the spectrum of the light coming from different angles is different (e.g. sun and sky) not only the irradiance measured may be inaccurate but also the shape of the spectrum may be distorted. When measuring outdoors, the size of this error will change through a day as the sun moves across the sky. The very small cosine diffusers sold by the spectrometer manufacturers tend to be prone to large errors, and individually calibrated, high quality diffusers like the D7-SMA and D7-H-SMA from Bentham (see section 3.20 on page 116 for full address) are preferable, although they are much more expensive (Figure 3.20).

---

[7]The measured biologically effective irradiance can be as large as twice its true value.

Figure 3.20: Bentham D7-H-SMA cosine diffuser. Photograph: Pedro J. Aphalo.

3.7.2.2 Calibration and corrections

When measuring UV-B with an array spectroradiometer it is not enough to have it properly calibrated, its optical characteristics (slit function at different wavelengths, stray light properties) need to be measured and a correction algorithm developed and later applied to each measurement. This makes the use of array spectroradiometers for characterizing UV-B doses complicated and error prone. This type of use has to be attempted only by experienced operators and the correction algorithm itself requires lots of effort to develop and implement. Given the lack of standardized procedures for stray light correction, its implementation requires advanced knowledge of optics and metrology. We will first discuss spectral calibration and thereafter stray light correction procedures.

Spectral calibration against standard lamps needs to be repeated regularly. For measurements not requiring very high precision, annual re-calibrations may be enough. However, the main consideration should be how valuable is the data that will be acquired. If the spectral sensitivity of the instrument has changed significantly from one calibration to the next, the data from all measurements done in between these calibrations are suspect, and should be discarded. Consequently if one does yearly re-calibrations one can lose one year's worth of data, while if one does monthly re-calibrations one only risks losing one month's worth of data. Consequently, the decision on how frequently to calibrate should, in addition to instrument stability, be based on the maximum size of the tolerable errors and on the value of the data (i.e. the cost of replacing the data if they need to be discarded).

The most common and stable calibration light sources are incandescent lamps (e.g. FEL lamps) with electronics in the power source which keeps the electrical power at the filament constant within very narrow margins. The distance between the lamp and the entrance optics, and their alignment, should also fall within a very narrow margin of the expected values. Calibration lamps are secondary or tertiary standards, connected by a chain of calibration steps to a standard kept at a metrology agency like NIST. Calibration lamps are supplied with spectral data about their emission characteristics. Calibration of the instrument is done by measuring the known spectrum and irradiance of the calibration lamp. Of course the output of the lamp will not exactly match the data supplied with it, because its original calibration is also subject to errors. Furthermore, there are errors deriving from slight differences between the burning conditions (current and voltage) during measurements and those when it was calibrated at the factory. Further errors can be introduced by small differences in the geometry of the optical setup. So, do not forget that calibrations are subject to errors. Furthermore, you cannot obtain an absolute estimate of calibration errors by comparing two instruments calibrated with the same lamp, unless this lamp is the primary standard.

Calibrating a spectroradiometer in the UV-B band with a FEL lamp is not recommended, because FEL lamps emit very little UV-B. For calibration in the UV-B deuterium lamps need to be used. Irradiance emitted by deuterium lamps is less stable than that emitted by FEL lamps. For coarse calibration the use of a deuterium lamp may be enough, but for accurate calibration it is best to use FEL and deuterium lamps in tandem. The shape of the spectrum emitted by deuterium lamps is stable, by matching the irradiances at wavelengths where the emission of both types of lamps overlap, one can extend an accurate calibration to shorter wavelengths. Spectrometer manufacturers also sell calibration light sources (lamp plus electronics) that may be good for routine calibration or especially for checking that calibrations performed in an optical bench remain valid. Again, what type of calibration procedure and lamp to use will depend on the accuracy required. If we want our measurements to be within $\pm 10\%$ of the true value we will need to use very good equipment and protocols for the calibration. If we can tolerate errors of, for example, $\pm 25\%$, calibrations can be less accurate.

It is also very important to do a wavelength calibration and to check this calibration regularly. It should not be forgotten when doing this calibration that it is affected by the temperature of the instrument as temperature affects (by thermal expansion) the dimensions of the optical bench and its components. Wavelength calibration is done based on elemental emission lines in discharge lamps (or even the sun). For quick checks low pressure mercury or germicidal lamps may be used. The manufacturers of spectrometers also sell special light sources for wavelength calibration. One should choose carefully which wavelengths to use (for example 253.652 nm, 296.728 nm, 334.148 nm, and 404.657 nm for mercury lamps, as they are simple peaks rather than multiple peaks very close together like those at 302 nm, 313 nm and 365 nm). If one desires a calibration accurate to a fraction of the wavelength step of the array one needs to fit a bell-shaped curve to the pixel showing the highest signal and those adjacent to it, to find the true location of the peak centre, most likely in-between two pixels.

To keep errors within ±10% in the UV-B when measuring sunlight, and especially to keep errors within ±10% for biologically effective doses, a good calibration is not enough when using single monochromator spectroradiometers. There is also a need to correct for stray light. If we do not correct for stray light some biologically effective doses will be overestimated by more than 100%. The ratio between stray light in the UV-B band and the maximum spectral irradiance measured in good single monochromator spectrometers is approximately 1×10^{-3}, while in double monochromator spectroradiometers it can be as low as 1×10^{-6}. If time for scanning, cost and lack of portability are no obstacles, it's preferable to use double monochromator instruments and these should also be used as the main instrument in a laboratory.

When applying the stray light correction, a thorough characterization of the slit function at different wavelengths and a check of the wavelength dependence of stray light are needed. This characterization does not need to be repeated, unless changes are made to the optical bench of the instrument. So, in contrast to the spectral calibration, the stray light characterization needs to be performed only once during the lifetime of the instrument unless major repairs or modifications are made.

In some array spectrometers, depending on the configuration, the width of the slit function may vary with the wavelength. This can introduce errors that are very difficult to correct. In some cases it might be preferable to chose a grating giving the instrument a relatively narrow wavelength range, for example 250 nm to 500 nm if the intended use is to measure UV irradiance.

The use and calibration of array spectroradiometers for measurement of UV-B radiation in sunlight is discussed in detail in the WMO report by Seckmeyer, Bais, Bernhard, Blumthaler, Drüke et al. (2010). Stray light correction methods are discussed in the papers by Ylianttila et al. (2005), Coleman et al. (2008), and Kreuter and Blumthaler (2009) and the references therein.

3.8 Underwater sensors and profiling

3.8.1 Measuring underwater UV radiation

Measuring UV radiation in the aquatic environment is difficult. Waterproof UV-measuring devices are needed or sensors protected in water proof housings as well as a means of deploying the sensors at the desired depth. Frequently, underwater measurements are referenced to the (spectral) irradiance at the surface of the water body, measured simultaneously with a matched "atmospheric" or terrestrial sensor.

As was mentioned in section 1.6 on page 15, measurement of the underwater UV field presents particular complexities, mainly related to variable attenuation occurring in different water bodies (Kirk, 1994a). Depending on the physical and chemical characteristics of the water, UV irradiance may decrease much more rapidly than PAR irradiance (this phenomenon is known as 'spectral leakage'), resulting in a changing spectrum with depth. Thus, underwater UV instruments are normally equipped with different filters and photodiodes to minimize these effects and to improve the sensitivity to particular wavelengths of interest. As discussed above in section 3.5.4 on page 83, errors are introduced if the spectrum being measured differs from that of the source used for calibration of the broadband instrument. This implies that when using broadband sensors underwater the errors will depend on the depth at which the radiation is being measured. In addition, a general problem exists for broadband sensors and spectrometers if sensors are only calibrated in air. The same calibration function cannot be used with the sensor in air and in water. Wavelength-dependent correction factors, so called immersion factors, must be used to adjust the signal if sensors are immersed in water as for example described by Ohde and Siegel, 2003.

Water movement and weather conditions can affect the measurements of underwater UV radiation. For an accurate determination of UV radiation in the field, sunny, cloudless conditions and calm waters are preferred. The effect of waves may cause difficulties especially when measuring near the surface. On the other hand, clouds and other atmospheric UV absorbing phenomena can alter the conditions during the measurement of vertical light profiles, especially when the diffuse component of the beam is altered.

Solar zenith angle is an important factor which affects the irradiance above the water surface, and also the reflectance of the water surface for the wavelengths of biological relevance. In addition, the radiation amplification factor (RAF, see also section 3.10), can be used to estimate the effect of changes in the ozone column on biologically effective exposure. Corrections can be applied to measurements performed with broadband instruments, but measurements with narrow-band multi-filter instruments and spectroradiometers are less error prone.

If our objective is to describe the characteristics of the waters, UV radiation should be measured around solar noon, when solar elevation is maximal. However, if we are interested in describing the daily UV exposure of some organism, measurements should be done preferably during most of the day and at the depths of interest.

3.8.2 Profiling

Profiling is the measurement of irradiance or spectral irradiance as a function of depth in a water body. Special frames are used for lowering the instruments through the water. For light measuring instruments, the frame or rig should not occlude the field of view of the radiation sensors. A means of determining the depth at which the sensor is located, and any deviation from a vertical orientation should also be available. Suitable cabling is used to connect the underwater sensors to onboard computers or dataloggers.

3.8.3 Underwater radiometers

Various underwater radiometers are currently available and their accuracy and characteristics vary considerably. The most appropriate instrument to choose depends on the specific goals of a study. Different types of radiometers include broadband radiometers, narrow-band multifilter radiometers, photodiode array spectroradiometers, and scanning spectroradiometers, including single monochromator spectroradiometers. Comparisons of the main characteristics of different types of instruments have been published (S. B. Díaz et al., 2000; Kirk, 1994a; Kjeldstad et al., 2003; Tedetti and Sempére, 2006). Some commonly-used underwater UV instruments are described below.

The **ELDONET radiometers** (Real Time Computer, Germany) were developed within the framework of the European light dosimeter network (Figure 3.21) and have been described in detail (Häder et al., 1999). The dosimeters are three-channel broadband filter devices with an entrance optic consisting of an integrating Ulbricht sphere with an internal $BaSO_4$ coating (Khanh and Dähn,

1988). Silicon photodiodes (BPX60 for the PAR range and SFH291 for the two UV wavebands, both from Siemens, Germany) are used in combination with custom-made filters to select the wavelength ranges for UV-B (280–315 nm), UV-A (315–400 nm) and PAR (400–700 nm), a custom-made interference filter for UV-B (Janos Technology, Townshend, VT, USA), a DUG 11 band filter for UV-A (Schott & Gen., Mainz, Germany) and a broad band filter for PAR (WBHM, Optical Coating Laboratory, Santa Rosa, CA, USA). Eldonet performs 60 measurements each minute.

Submergible multichannel radiometers like PUV 500, PUV 2500 series from Biospherical Instruments (San Diego, USA) are equipped with narrow-band filter detectors in the range of UV and PAR (Figure 3.22). These radiometers are equipped with depth and temperature sensors and thus are well suited for accurate light profiling. The spectral characteristics of the five filters used in the PUV 500 instrument are as follows: 305 ± 1 nm (band pass 7 ± 1 nm) 320 ± 2 nm (band pass 11 ± 1 nm) 340 ± 2 nm (band pass 10 ± 1 nm) 380 ± 2 nm (band pass 10 ± 1 nm)

On the other hand, the PUV 2500 measures 7 (optionally 8) wavebands of downwelling irradiance (305, 313, 320, 340, 380, 395 nm and PAR: 400–700 nm) with one upwelling radiance channel (natural fluorescence). Each channel with 10 nm FWHHM except 305 (controlled by atmospheric ozone cutoff). The instrument includes pressure/depth sensing (350 m maximum) and temperature control. A 32 channel multiplexer selects signals from 8 photodetectors, temperature, pressure and tilt/yaw detector. The cosine collector is made of Teflon-covered quartz for use in the water.

The RAMSES family of **hyperspectral radiometers** (Trios GmbH, Germany) are miniature single monochromator spectrometers with a resolution of 2 to 3 nm per pixel and 100 or 190 usable channels in the photodiode array (Figure 3.23). They can be used in air or in water. The Ramses ACC-UV is an integrated UV hyperspectral radiometer, and the Ramses ACC-VIS is a UV-A and visible hyperspectral radiometer, both equipped with a cosine collector. Ramses ASC-VIS is equipped with a spherical collector shielded so as to measure radiation from one hemisphere. To measure scalar irradiance, two of these sensors can be deployed pointing in opposite directions. They are calibrated for underwater and air measurements (two different calibrations). The device has a small size, the signal capture requires some power consumption and portable (laptop) terminal at the surface. The detector type is a silicon photodiode array designed to capture wavelengths between 320–950 nm for VIS models and 280–500 nm for the UV models, with an irradiance accuracy better than 6-10% depending on the spectral range.

Although the LI-1800UW instrument is not currently

Figure 3.21: Underwater Eldonet (European light dosimeter network). (Photograph by Donat Häder, reproduced with permission.)

Figure 3.22: Biospherical PUV500 and PUV 510 (underwater). Left: © Biospherical Instruments, Right: Félix López Figueroa.

Figure 3.23: RAMSES-ACC-UV hyperspectral radiometer. © Trios GmbH.

produced by LI-COR, many foundational studies focused on UV penetration during the 1980s and 1990s were carried out using this spectroradiometer. The optics of this scanning spectroradiometer is based on a single holographic monochromator grating, a silicon detector and a filter wheel to improve stray light rejection. The wavelength range is between 300 and 850 nm, with a bandwidth of 8 nm and accuracy of ± 1.5 nm. Originally optional slits of different widths were available, so these specifications vary with the exact configuration used. The whole optical bench and the microcomputer system is contained in the massive waterproof housing designed for measurements to a depth close to 200 m. Being a single monochromator instruments its accuracy is limited by stray light when used to measure UV-B radiation in daylight.

The OL 754-O-PMT Spectrometer Optics Head (Optronic Laboratories) is based on a double monochromator for low stray light and measuring from 300 to 850 nm. Configurations with other gratings giving different wavelength ranges are available. The system utilizes holographic gratings with peak efficiencies at 300 nm. The instrument can be fitted with an OL IS-470-WP Submersible Sphere Assembly (4-inch integrating sphere) attached by means of quartz optical fibre to the non-submersible spectroradiometer. The sphere follows a dual port design with an entrance port and an exit port located 90° apart. The sphere contains an internal baffle in front of the exit port to permit only light reflected by internal surface of the sphere to exit the sphere and enter the fibre.

Another approach to measuring underwater UV radiation is to protect a regular UV sensor, as for example, those described in section 3.5.2 on page 82, within a hermetic water-proof housing (Figure 3.24). Of course the enclosure should have an UV transparent window, and the sensors must be calibrated inside the enclosure.

3.9 Modelling

For locations and times not covered by measurements, alternative approaches have to be considered for estimating the prevailing radiation conditions. For this purpose, various methods for modelling the surface UV radiation have been developed. These range from simple theoretical-empirical methods for estimating the clear-sky surface radiation to more sophisticated methods that account also for the effects of clouds as inferred either from ground-based station data or satellite measurements. Table 3.3 gives a simplified view of the main features of methods that are available for modelling the surface UV radiation.

Simple theoretical-empirical methods, such as those of Björn and Murphy (1985) and Bird and Riordan (1986),

have been widely used thanks to their fairly simple user interface. These methods provide spectral surface irradiances, optionally on tilted surfaces, and account for the main parameters affecting the surface radiation conditions under cloudless skies. Results indicate that they predict the surface radiation with reasonable accuracy as compared to more detailed radiative transfer calculations and measurements (Bird and Riordan, 1986).

Another user-friendly alternative is to use an interactive web-based interface to radiative transfer simulations, such as the FastRT (Engelsen and Kylling, 2005, available at `http : / / nadir . nilu . no / ~olaeng / fastrt / fastrt . html`) or the QUICK TUV (`http://cprm.acd.ucar.edu/Models/TUV/Interactive_TUV/`). Both of these are based on rigorous radiative transfer models, which means that their accuracy depends mainly on the choice of values for the input parameters. Both include a selection of biological weighting functions, and FastRT also provides the possibility to account for the effect of clouds.

When cloud effects need to be accounted for in detail, either satellite methods or so-called UV reconstruction methods should be considered. The UV reconstruction methods usually rely on ground-based measurements of some kind for accounting for the effect of clouds on radiation. Although the methods vary in their exact approach, the idea of all of them is to utilize available observations to account for the parameters that affect the amount of UV radiation reaching the surface. These parameters are, most importantly, clouds, total ozone column, surface albedo, atmospheric aerosols, and altitude or pressure. In addition, the solar zenith angle determines the path length of the direct radiation through the atmosphere and is therefore the single most important factor for the surface UV radiation. The Earth-Sun distance, which varies over the course of the year, also needs to be accounted for.

Many station-based methods for reconstructing the surface UV radiation were included in recent European efforts to gain a better understanding of past UV radiation and its climatological behaviour (den Outer et al., 2010; Koepke et al., 2006). Similar methods have been proposed and applied in other parts of the world as well (V. Fioletov et al., 2001; V. E. Fioletov et al., 2004). Compared to satellite methods, the advantage of the station-based methods is that they tend to give more accurate estimates of the surface UV radiation. In particular, methods using global solar radiation (300–3000 nm), measured by pyranometers at numerous stations worldwide, as input for determining the cloud effect typically show small bias and fairly small scatter when compared to measurements (Koepke et al., 2006). To mention one

Table 3.3: Simplified properties of methods for modelling of surface UV radiation

METHOD	SPECTRAL OR WEIGHTED	CLOUDS IN-CLUDED	COMMENTS	EXAMPLE REFERENCE
Simple theoretical-empirical	both	no	user friendly; reasonable accuracy for cloud-free atmospheres	Björn and Murphy, 1985
Interactive radiative transfer	both	yes/no	user friendly; based on rigourous radiative transfer calculations	Engelsen and Kylling, 2005
Pyranometer-based UV reconstruction	mostly selected weighting	yes	good performance compared to measurements; requires expert to use	den Outer et al., 2010
Satellite	mostly selected weighting	yes	tendency toward overestimation; easy access to data and global coverage	Kujanpää et al., 2010

Figure 3.24: Hermetic water -proof housing without and with light sensors and data logger inside of the box. Photographs © EIC (Equipos, Instrumentación y Control).

example, Lindfors et al. (2007) estimated daily erythemal UV doses at four Nordic stations, and found a systematic difference of between 0 and 4%, depending on the station, and a root-mean-square error of 5–9% as compared to measurements for the summer period.

Satellite methods are based on radiative transfer simulations combined with information on, for example, clouds and total ozone column retrieved from the satellite observations. Satellite retrieved UV irradiances typically show an overestimation of 10% or more, and a large scatter, when compared to surface measurements: the root-mean-square errors for daily erythemal UV doses tends to be of the order of 30–40% (Kujanpää et al., 2010; Lindfors, Tanskanen et al., 2009; Tanskanen et al., 2007). The main part of the overestimation is usually attributed to aerosol absorption, which is not accounted for properly in current satellite UV algorithms. The advantage of satellite methods, on the other hand, is their large geographical coverage, often global, and easy access to data that they provide.

Most methods, both station-based UV reconstruction methods and satellite methods, only provide UV data corresponding to a selection of weighting functions (e.g., erythemally weighted UV), and, in addition, sometimes irradiances for selected wavelengths. Furthermore, the methods typically include only irradiances for a horizontal surface. This may become an obstacle for biological applications where, for example, a specific weighting function or spectral information would be preferred. In principle, however, many of the methods could be extended to produce spectral irradiances and fluence rates. On the other hand, fluence rates can also be estimated based on the horizontal irradiance (e.g., Kylling et al., 2003).

Recently, Lindfors, Heikkilä et al. (2009) presented a method for modelling spectral surface irradiances. The method relies on radiative transfer simulations, and takes as input (1) the effective cloud optical depth as inferred from pyranometer measurements of global radiation; (2) the total ozone column; (3) the surface albedo as estimated from measurements of snow depth; (4) the total water vapour column; and (5) the altitude of the location.

Figure 3.25 shows a comparison between the daily accumulated irradiances at 300 and 320 nm from this method and measurements with a Brewer spectroradiometer at Jokioinen, southern Finland. At both wavelengths, the reconstructed irradiances closely follow the measured ones.

A variety of methods for modelling the surface UV radiation are available. Which method is the best, or the most appropraite, depends on the specific question that is to be answered. In general, the complexity of the method tends to grow with increasing accuracy, and the pyranometer-based UV reconstruction methods, that are considered to provide highest accuracy, typically require an expert user. The use of such a method will, however, increase cross-disciplinary collaboration and may therefore be worthwhile.

3.10 BSWFs and effective UV doses

The emission spectrum of UV-B lamps, even filtered with acetate, is different to that of the effect of ozone depletion. The spectrum of the effect of ozone depletion not only changes with ozone column thickness, but also with solar elevation. In other words, it changes through a day and with seasons (Figure 3.26). Because of this, it is almost impossible to exactly simulate the effect of ozone depletion in field experiments. The best we can do is to calculate effective doses.

Biologically effective exposures (see Box 3.1) are a way of measuring radiation differing in spectral composition with the same 'measuring stick'. This 'measuring stick' is a biological response. Behind each UV-B_{BE} measurement we need to assume the involvement of a biological response. If we know the action spectrum for the biological response, we can use it as a BSWF: We multiply, wavelength by wavelength the spectral irradiance of the light source by the BSWF obtaining a weighted spectrum, giving a biologically-effective *spectral* irradiance (Figure 3.27). We then integrate the result over wavelengths, to obtain a single number, the biologically effective UV irra-

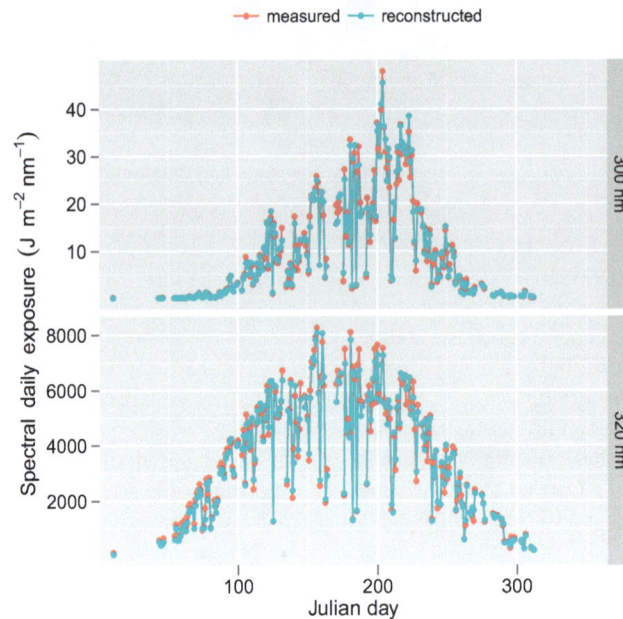

Figure 3.25: Reconstructed and measured daily cumulative irradiances at (a) 300 nm and (b) 320 nm at Jokioinen during 2004. From Lindfors, Heikkilä et al., 2009.

Box 3.1: Biologically effective irradiance: what is it, and what it is used for?

By 'biologically effective irradiance' we mean the irradiance weighted according to the effectiveness of different wavelengths in eliciting a photobiological response. The most frequently used biologically effective irradiance quantities are photometric quantities such as those described in Box 1.1 on page 6. In the case of photometric quantities the (energy) irradiance is weighted according to the response of the human eye.

When studying the effects of UV-B radiation on plants we use as biological spectral weighting functions (BSWFs) spectra describing the response of some plant function. For example an action spectrum for accumulation of flavonoids, or an action spectrum for growth inhibition. To be able to calculate biologically effective irradiances using any BSWF, we need to measure the spectral irradiance of the light source.

If we integrate the effective irradiance for the duration of an experiment then we obtain a biologically effective exposure (usually called 'biologically effective dose' by biologists). If we do the time integration for one day we obtain a biologically effective daily exposure.

These weighted irradiances are usually expressed using units corresponding to the underlying energy irradiances, independently of the BSWF used. Quantities calculated using different BSWFs are expressed in the same units, but the values cannot be compared because in reality they are measured on different scales.

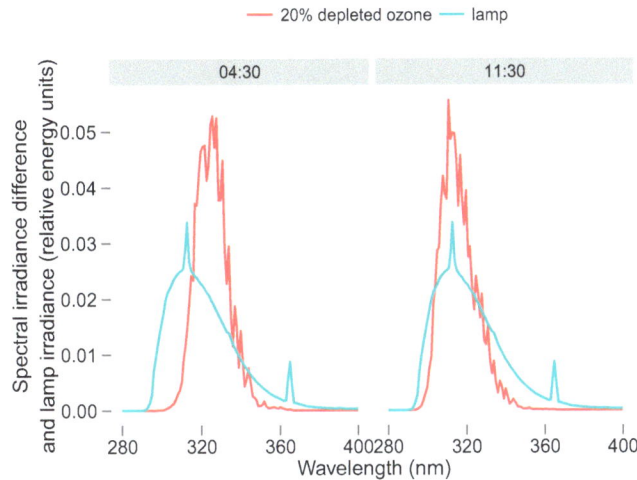

Figure 3.26: Comparison of spectral (energy) irradiance from Q-Panel UVB-313 lamps filtered with cellulose acetate film and the effect of 20% ozone depletion on the solar spectrum at two times of the day. For details of the simulations see fig. 1.7. Spectral irradiances normalized to equal total UV energy irradiance (= area below the curves).

diance[8]. In Figure 3.27 it also possible to appreciate the difference in relative change of this effective irradiance, for a given level of ozone depletion, depending on the BSWF used.

It is important to make sure that both the irradiance and the effectiveness are measured using compatible units. It is common to express action spectra as spectral *quantum* effectiveness and to measure light sources as spectral (*energy*) irradiance. In UV research, biologically effective doses are most frequently expressed in weighted energy units and to be able to calculate these doses from spectral (energy) irradiance measurements for a light source (sunlight or lamps) one needs to use an action spectrum expressed in energy effectiveness. Most common formulations of action spectra need to be transformed from quantum effectiveness to energy effectiveness (one important exception is the CIE erythemal spectrum formulation). This is something that is often neglected, and is a source of difficulties when comparing doses between different publications.

Another possibility for measuring effective doses is to have a broadband sensor with a spectral response resembling the action spectrum of interest. In practice the spectral response of such sensors is only an approximation to the desired BSWF and consequently need to be calibrated under the light source to be measured, usually by comparison to a double monochromator scanning spectroradiometer (see section 3.7 on page 85). Most such sensors are calibrated for sunlight, and consequently give

biased readings when used for measuring radiation from most lamps.

The radiation amplification factor (RAF) gives the percent change in effective UV dose (H) or, strictly speaking, effective UV exposure (H) for each percent change in ozone column thickness. It should be calculated using logarithms.

$$\mathrm{RAF} = \frac{\ln H^{\mathrm{d}} - \ln H^{\mathrm{n}}}{\ln[O_3]^{\mathrm{n}} - \ln[O_3]^{\mathrm{d}}} \qquad (3.6)$$

where H is dose and $[O_3]$ ozone concentration, and superscript d indicates ozone depleted condition, and superscript n indicates normal, or reference, ozone depth condition. The value of RAF depends strongly on the action spectrum used to calculate the effective dose. Looking at fig. 3.27, it can be understood why RAF is much larger for GEN than for PG.

3.10.1 Weighting scales

Ultraviolet action spectra are usually normalized to quantum effectiveness = 1 at 300 nm. This is arbitrary, and especially in the older literature, you will find action spectra normalized at other wavelengths. In the Materials and Methods section always report the normalization used, in addition to the action spectrum used as a BSWF. Remember that as the wavelength used for normalization is arbitrary, values of effective UV doses calculated using the same BSWF but normalized at different wavelengths

---

[8]Graphically this is the area under the curve.

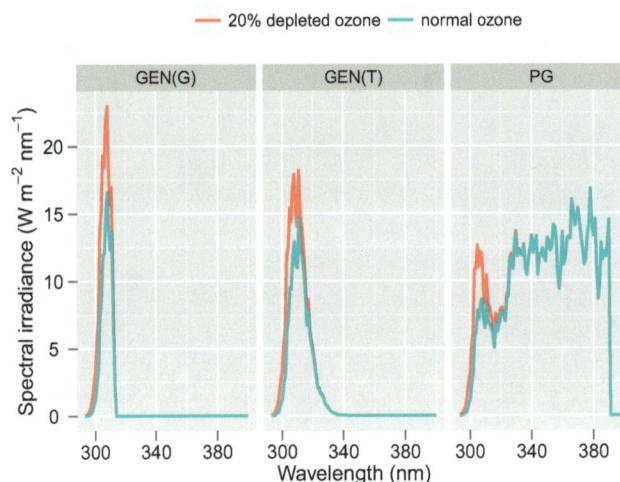

Figure 3.27: Biologically effective spectral irradiance. Solar UV energy spectrum weighted with the plant growth (PG) and generalized plant action spectra (GEN(G): Green's formulation; GEN(T): Thimijan's formulation)

cannot be directly compared because they are expressed on different scales. Of course biologically effective irradiances based on different BSWFs cannot be compared to each other either.

Using the correct BSWF is very important, as using the wrong BSWF has usually serious implications on the interpretation of experimental results (Caldwell and Flint, 2006; Kotilainen et al., 2011).

3.10.2 Comparing lamps and solar radiation

Frequently we want to compare UV doses in growth chambers to doses outdoors. Unless we have a solar simulator the spectra will differ significantly. It is especially important to keep the UV-B:UV-A:PAR ratio similar to that in solar radiation. Table 3.4 gives an example comparing a frame with two acetate-filtered Q-Panel UVB-313 lamps to sunlight with normal and 20% depleted ozone at Jokioinen for 21 May at 11:30.

3.10.3 Effective doses, enhancement errors and UV-B supplementation

Ultraviolet-radiation supplementation can be modulated so as to follow natural variation in solar UV or just follow a daily square wave pattern (see section 2.2.6 on page 48). It is important that lamps are filtered with cellulose di-acetate film (to remove UV-C radiation emitted by UV-B lamps, which is absent in natural sunlight) and that these filters are replaced regularly, specially in the

case of square wave systems, as modulated systems with feedback compensate for the reduced dose automatically (although not for the change in spectrum) (Newsham et al., 1996). Even when adequately filtered the emission spectrum of UV-B lamps does not match the effect of O_3 depletion (Figure 3.28. We need to calculate effective doses using BSWFs.

Errors caused by the mismatch between the doses aimed at and those achieved when simulating the effect of ozone depletion with UV-B enhancement with lamps are called enhancement errors. The main source of these errors is the mismatch between the assumed spectral response and the real spectral response. This is so because the adjustment of the burning time (or power) of the UV-B lamps used for enhancement needs to be based on biologically effective doses. However, depending on the different BSWFs used, the needed lamp burning time may be long or short (Figure 3.29). Another way of looking at this problem is to compare the deviation of the achieved UV^{BE} when using a 'wrong'[9] BSWF compared to the target one—e.g. corresponding to a certain magnitude of ozone depletion. Figure 3.30 shows this comparison for the frequent case of use of a CIE-weighted broadband sensor to control the lamps used in experiments with plants. The errors are surprisingly small for GEN and PG.

An additional source of errors is shading by the lamp frames. If we do not attempt to compensate for the shade with UV-B from the lamps the error between PG and GEN is small (Kotilainen et al., 2011). How much

---

[9]By 'wrong' BSWF we mean one describing the spectral response of a different response than that under study.

Table 3.4: Effective UV-B irradiances. Lamps are Q-Panel UVB-313 tubes filtered with cellulose diacetate. Solar spectra are simulated for 21 May at Jokioinen, Southern Finland under clear sky conditions at 11:30 solar time. RAF is calculated from the doses for normal and 20% ozone depleted solar radiation, and indicates the percent change in biologically effective UV dose, for a 1% change in ozone concentration. See table 1.4 on page 25 for the key to the codes for the action spectra used as BSWFs and the references to original sources, and section 3.17 on page 111 for details on the formulations used for the BSWFs.

Source and units	GEN(G)	GEN(T)	PG	DNA(N)
Solar, normal O_3 (W m^{-2})	0.13	0.19	0.99	0.050
Solar, 20% depleted O_3 (W m^{-2})	0.20	0.26	1.04	0.081
Two lamps, at 40 cm (W m^{-2})	0.53	0.55	0.51	0.47
Two lamps, (% of solar normal)	408	290	52	940
Two lamps, (% of solar depleted)	265	212	48	580
RAF	1.84	1.31	0.24	2.10

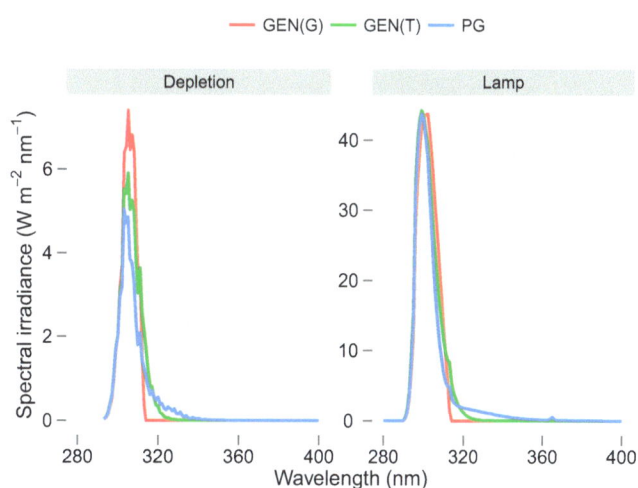

Figure 3.28: Effective spectral irradiances for UV-B lamps and the effect of 20% ozone depletion, based on GPAS (GEN(G): Green's formulation; GEN(T): Thimijan's formulation) and PG action spectra.

Table 3.5: Lamp burning times in minutes needed to compensate for 10% shading throughout a day for eight different BSWFs. The calculations are based on simulated solar spectra for 21 May and 21 June at Jokioinen, southern Finland, under clear sky conditions. The action spectra are in order of increasing burning time from left to right. See appendix 3.17 for details about the spectra. Adapted from Kotilainen et al. (2011)

BSWF	DNA(N)	GEN(G)	GEN(T)	CIE	FLAV	DNA(P)	PG	PHIN
May	4.2	9.4	15.3	17.0	29.1	52.8	119	417
June	5.7	12.7	19.3	20.8	35.0	61.7	133	465

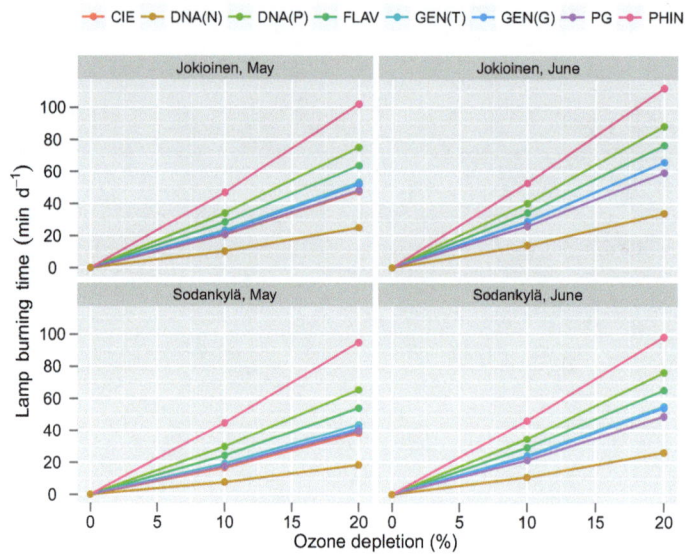

Figure 3.29: Lamp burning times needed to achieve an effective UV dose equivalent to that under the target level of ozone depletion. Calculations based on solar spectra simulated for clear sky conditions for two locations in Finland, for 21 May and 21 June. Redrawn from Kotilainen et al. (2011).

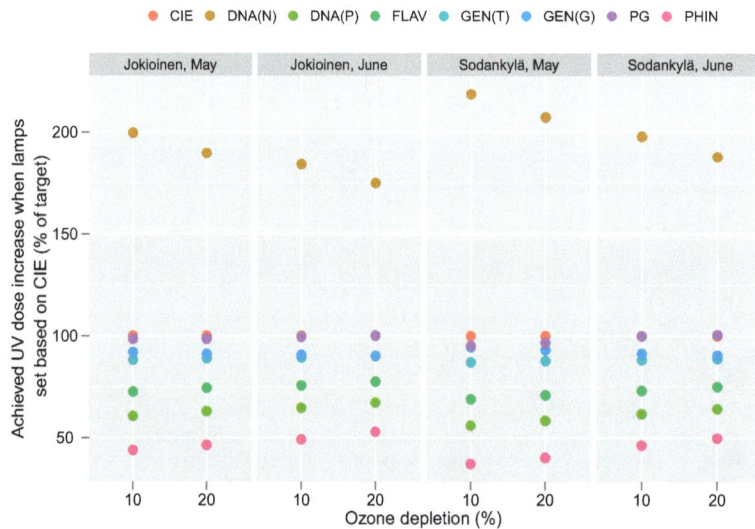

Figure 3.30: Achieved enhancement assuming different action spectra as the true one, with lamp burning times set based on using the CIE action spectrum as BSWF. Calculations based on solar spectra simulated for clear sky conditions for two locations in Finland, for 21 May and 21 June. Redrawn from Kotilainen et al. (2011).

the effective UV dose decreases with shading does not depend on the BSWF used as long as the shade is 'gray' (equally affects all relevant wavelengths). However, how much UV from the lamps will be needed to compensate for this will depend on the BSWF used (Table 3.5). With some BSWFs the lamp power or burning time needed to compensate for shading is much more than that needed to simulate ozone depletion, because with spectra like PG, we need to replace shaded UV-A with UV-B from lamps. See section 3.10 on page 101 for a discussion of the relationship between changes in ozone column and changes in effective UV radiation.

To minimize shading errors we must build lamp frames that produce little shade. Probably < 5% shading is achievable. We should not attempt to compensate for shading by the frames with UV from lamps. Shifting the whole experiment's UV baseline dose by a small percentage but keeping the size of enhancement at the target level is probably the best approach available.

3.11 Effective UV doses outdoors: seasonal and latitudinal variation

The state of the atmosphere (in terms of ozone column thickness, cloudiness and aerosol content) together with day-length and daily course of the solar zenith angle (θ) are the main factors determining the climatology of UV (and its components, UV-B and UV-A) at ground level. For clear sky conditions the average spatial and temporal distribution of UV irradiance can be computed by means of a radiative transfer model fed with proper O_3 and aerosol data (Grifoni et al., 2009, 2008). In these conditions the latitudinal and temporal distribution of UV irradiance are driven mainly by two factors: day-length and θ through the day. See section 3.9 on page 97 for a discussion of different approaches to modelling.

To go from the climatology of UV spectral irradiance to that of the biologically effective UV (UV^{BE}) exposures (or doses), spectral irradiance has to be weighted to account for the different efficiency of each wavelength in producing biological effects; this is done applying a BSWF—based on the action spectrum of the biological process considered. Plant action spectra differ in the weight given to UV-B and UV-A radiation, as it has been illustrated in section 1.8 on page 21. In this analysis Grifoni et al. (2009) considered two action spectra related to plant response: the so-called Generalized Plant Action spectrum (GEN, proposed by Caldwell, 1971) and the more recent Plant Growth spectrum (PG, Flint and Caldwell, 2003). The erythemal action spectrum (CIE, McKinlay and Diffey, 1987) was also included since instruments with a spectral sensibility quite close to it have been used in several field experiments. The analysis presented in this section was based on spectral global irradiance for cloud-free conditions on horizontal surfaces simulated by means of the STAR model (Ruggaber et al., 1994; Schwander et al., 2000) for Rome, Italy (lat. 41°.88 N, long. 12°.47 E), Potsdam, Germany (lat. 52°.40 N, long. 13°.03 E) and Trondheim, Norway (lat. 63°.42 N, long. 10°.42 E) on the first day of each month of the year from 1:00 to 23:00 (UTC time) with a 30 min time step. Daily UV^{BE} doses ($kJ\,m^{-2}\,d^{-1}$) were calculated for all these 12 days.

After convoluting the UV spectral energy irradiances at ground level for the three locations and seasons with the three BSWFs, a picture of the seasonal of UV^{BE} radiation was obtained, which was different depending on latitude. The pattern of seasonal and latitudinal variation of the daily UV^{BE} doses is shown to depend strongly on the BSWF, and day of the year (Table 3.6).

The changes in daily UV^{BE} doses occurring through the year relative to the yearly average, for the three locations across Europe illustrate the effect of action spectra used as BSWFs (Figure 3.31).

The largest seasonal variation occurs when the UV^{BE} daily doses are computed on the basis of action spectra completely or partially excluding the contribution of the UV-A component, as in the case of GEN and CIE respectively. Figure 3.31 shows the extent to which latitude affects these seasonal changes, which have a larger amplitude as the latitude increases. In other words, the climatology of UV^{BE} appears to be strongly dependent on the action spectrum used, and, as different plant responses follow different action spectra, the effective UV climatology will depend on the plant response under study.

These differences in climatology are ecologically relevant, for instance for perennial species that may experience different seasonal change in the UV^{BE} irradiance they are exposed to, depending on the latitude at which they are growing and on the biological process/action spectrum studied.

3.12 Effective UV doses in controlled environments

If the light spectrum in a controlled environment differs greatly from that under natural conditions, the relative biologically effective irradiances will differ greatly depending on the action spectrum used as BSWF. For example, if UV-A irradiance is much lower in daylight than in the controlled environment, but UV-B irradiance similar to that in daylight, the dose calculated using GEN(G) as a BSWF will differ little between the controlled environment and daylight, but the dose based on PG will be much lower in the controlled environment than in daylight. This example shows why we need to use a real-

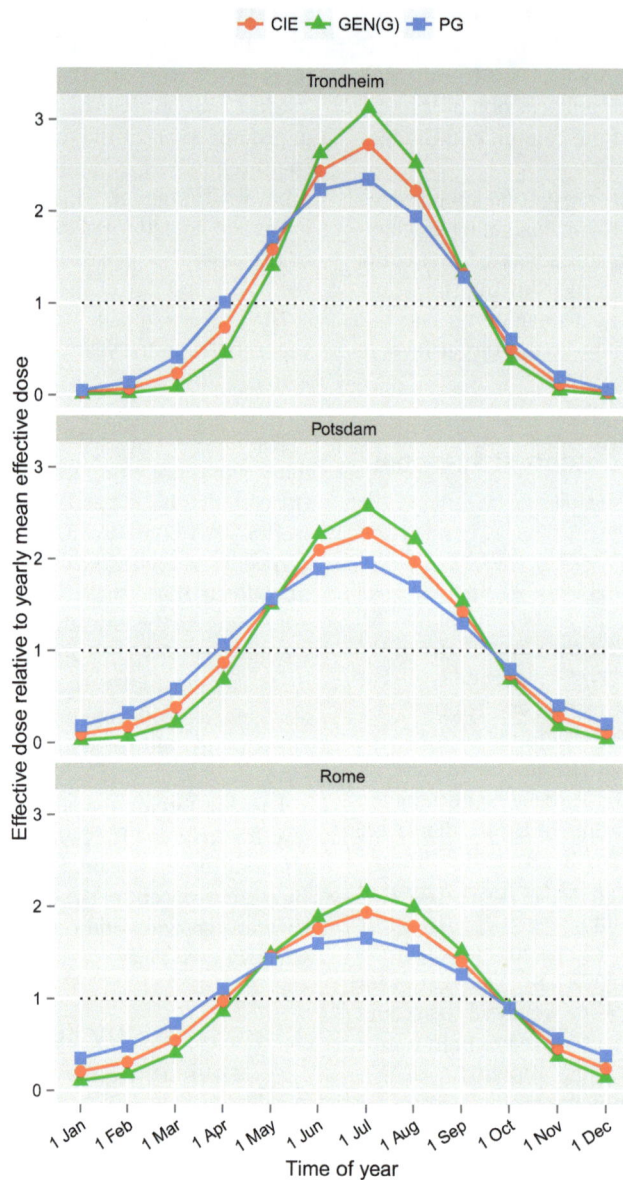

Figure 3.31: Relative seasonal variation of daily biologically effective UV doses for Trondheim, Potsdam and Rome, calculated using three different BSWFs. Actual values are normalized by division by the yearly average for each location. The horizontal dotted line shows the normalized mean. See Table 1.4 on page 25 for references to the sources of the spectra used as BSWFs.

Table 3.6: Seasonal and latitudinal variation in effective UV doses across Europe. Daily doses ($kJ\,m^{-2}\,d^{-1}$) at three times of the year at three locations. Modelled values for clear sky conditions. See text for details. See Table 1.4 on page 25 for references to the sources of the spectra used as BSWFs.

Date	City	CIE	GEN(G)	PG
1 Jan	Trondheim	0.028	0.008	0.65
1 Jan	Potsdam	0.161	0.047	3.25
1 Jan	Rome	0.515	0.296	7.77
1 Apr	Trondheim	0.92	0.528	13.87
1 Apr	Potsdam	1.56	1.27	18.78
1 Apr	Rome	2.46	2.48	24.70
1 Jul	Trondheim	3.41	3.67	32.30
1 Jul	Potsdam	4.07	4.77	34.60
1 Jul	Rome	4.90	6.23	37.08

istic light spectrum in controlled environments when we want to extrapolate the results from such experiments to natural conditions. Furthermore, an increased ratio between UV-B and UV-A radiation, and/or UV-B radiation and PAR artificially enhances the responses of plants to UV-B radiation.

3.13 An accuracy ranking of quantification methods

Table 3.7 shows a preliminary comparison of different methods available for quantifying solar UV radiation and estimating biologically effective doses. Bias is the systematic directional deviation from the true value, for example overestimation of irradiance or doses. Uncertainty is a random deviation that prevents us from knowing the true value, but deviations are not systematic —measurements from different instruments or the same instruments after different recalibrations will deviate by different amounts and in different directions from the true value.

A detailed comparison is difficult with the information currently available, consequently some gaps remain in the table. However, some general recommendations are possible. For outdoors experiments with manipulation of solar UV, the best option is most probably a combination of (a) hourly simulation of UV spectral irradiance with a model using ground station data as input, plus (b) spot measurements, for example under different filters with a spectroradiometer under clear sky conditions. It is best to replace modelling (a) with actual continuous measurements with a well calibrated double monochromator spectroradiometer. However, this is rarely possible in practice, as there are few ground stations producing validated spectral data.

For experiments using lamps, the best option from the point of view of accuracy is the use of a double mono-

chromator spectroradiometer. From a practical viewpoint, using a broadband instrument calibrated against a double monochromator spectroradiometer, may be easier. When measuring mixed daylight and lamp radiation, or mixed radiation from diverse lamps, or when there is degradation of filters, broadband instruments are totally unsuitable.

Single monochromator spectroradiometers should be avoided for measurements of UV-B radiation when there is a background of UV-A or visible radiation, unless very special handling of stray light is done by a combination of special measuring protocols and data processing.

3.14 Sanity checks for data and calculations

When quantifying UV radiation, or in fact when doing any measurement, one should compare the values (e.g. irradiances or daily doses) against what has earlier been reported in the literature for a similar light source. In this way many errors can be detected. To help in this process we present in Table 3.8 typical values for both unweighted UV-A and UV-B (energy) irradiance, photon irradiance, and biologically effective irradiances with the most frequently used BSWFs and wavelength normalisations. For UV radiation outdoors one can use a model (e.g. TUV quick simulator) to estimate spectral irradiance values and from them calculate effective doses or exposure. We have done these calculations for some sites, and present them in Table 3.8. We have also included data for some lamps.

3.15 | Recommendations |

In this section we list recommendations related to the quantification of UV radiation in experiments. See sections 2.6.1 and 2.6.2 for recommendations on manipu-

Table 3.7: Achievable accuracy for different methods used for quantifying UV radiation in sunlight. Most instruments commonly used in plant science research have even larger uncertainties due to suboptimal calibration and data-quality control. SR = spectroradiometer; TOMS = total ozone mapping spectrometer. θ: solar zenith angle.

Method	Effective-dose bias	Uncertainty
Single monochromator SR, no straylight correction	+50 to +150%	?[§]
Satellite (e.g. TOMS based UV)	>+10%	30–40%[†]
Dosimeters, erythemal	?	?[‡]
Modelling with ground station data	0 to +4%	5–9%[†]
Broadband sensors, erythemal, uncorrected	-15% to +15% depending on θ, ozone column and instrument	5–50%
Broadband sensors, erythemal, corrected for ozone column and θ	≈ 0	>10%
Single monochromator SR, with straylight correction	$\approx$+5 to +10%	?[§]
Double monochromator SR	very small, dependent on θ and instrument	5–10%

[†] for estimated daily erythemal doses compared to actual daily erythemal doses. [‡] for integrated erythemal dose over exposure period. [§] Uncertainties depend on environmental conditions, especially temperature at which the instrument is used, compared to its temperature during calibration.

lation of UV radiation, section 4.9 for recommendations about plant growing conditions, and section 5.12 for recommendations about statistical design of experiments.

1. Always report in your publications the UV radiation effective exposure (frequently called effective doses in the biological literature), depending on the type of experiment, as daily integrated values or effective irradiances plus daily exposure time. Calculate effective doses using relevant action spectra and if possible report doses using all the most commonly used ones: CIE, GEN(G), and PG.

2. Always indicate the name of the quantity measured, the type of sensor and its position during measurement, location of sensor with reference to plants, and unit of measurement. In the case of effective doses, indicate how they were calculated, in particular, cite the bibliographic source for the action spectrum and formulation used, and at which wavelength the spectrum was normalized to effectiveness equal to unity.

3. Whenever possible include in your publications the emission spectrum of the light source and/or transmittance spectrum of the filters used, or cite an earlier paper where the spectra have already been published.

4. Use only instruments with recent and valid calibration data for the measurements at hand.

5. Make sure, in the case of broadband sensors, that the calibration is valid for the light source being measured. For example, broadband UV sensors calibrated for sunlight should **not** be used for measuring irradiance under UV-B lamps.

6. If high precision is required in the measurements, apply all the necessary corrections. This is important both for broadband and spectral measurements.

7. Be aware of temperature effects on the functioning of the meter and sensor used and apply the required corrections or use temperature stabilized instruments.

8. Single monochromator array or scanning spectrometers can be used for measuring effective UV exposures in sunlight only with very serious limitations and only if complicated corrections are applied to the raw data to take into account stray light and the properties of the slit function. These corrections are not possible with the software provided by the makers of the instruments. Uncorrected measurements from this type of instruments are subject to huge uncertainties and, what is worse, bias. Use double monochromator spectroradiometers instead.

9. When using spectrometers configured with SMA connectors for optical fibres, do not detach the

Table 3.8: Typical irradiance values, in the visible, UV-A and UV-B wavebands for several natural and artificial sources of radiation. Daylight radiation spectra marked 'TUV' were simulated under clear sky conditions at midday for the longest and shortest day of the year at three latitudes. Typical ozone depth values were used. The simulations were done with the TUV quick calculator (http://cprm.acd.ucar.edu/Models/TUV/Interactive_TUV/). In the case of weighted irradiances not output by the TUV model, the resulting energy irradiance spectra were used as input for calculating the energy, photon and biologically effective irradiances with the UVcalc R package (see section 3.19 on page 113). The action spectra used as BSWFs are described in Table 1.4 on page 25. Lamp spectra are measured at 40 cm distance from a pair of lamps at room temperature, and driven at full power. TL12 are Philips 40W UV-B lamps, UVB-313 and UVA-340 are 40W Q-Panel fluorescent lamps. The spectra for Jokioinen (60°49′ N, 23°30′ E, Finland) were also simulated (see Kotilainen et al., 2011, for details). The irradiances for Helsinki (60°14′ N, 25°1′ E, Finland) were measured under the different filters under clear sky on 8 June 2011 near solar noon. A Maya Pro spectrometer was used and stray light and slit function corrections were applied (unpublished data of T. Matthew Robson).

Light source	energy irradiance			photon irradiance			biologically effective irradiance					photon ratios	
	PAR ($\frac{W}{m^2}$)	UV-A ($\frac{W}{m^2}$)	UV-B ($\frac{W}{m^2}$)	PAR ($\frac{\mu mol}{m^2 s}$)	UV-A ($\frac{\mu mol}{m^2 s}$)	UV-B ($\frac{\mu mol}{m^2 s}$)	GEN(G) ($\frac{W}{m^2}$)	GEN(T) ($\frac{W}{m^2}$)	PG ($\frac{W}{m^2}$)	DNA(N) ($\frac{W}{m^2}$)	CIE ($\frac{W}{m^2}$)	UV-B:UV-A (×1000)	UV-B:PAR (×1000)
TUV, 250 DU, 0° N, 21 Dec., s.l.	n.a.	60	2.0	n.a.	180	5.5	0.47	0.56	1.6	0.24	0.30	30	n.a.
TUV, 250 DU, 0° N, 21 Mar., s.l.	n.a.	65	2.5	n.a.	200	6.5	0.59	0.68	1.8	0.30	0.37	33	n.a.
TUV, 250 DU, 0° N, 21 Mar., 4 km a.s.l.	n.a.	74	3.1	n.a.	223	8.0	0.75	0.86	2.1	0.40	0.46	36	n.a.
TUV, 300 DU, 30° N, 21 Jun., s.l.	n.a.	62	2.0	n.a.	188	5.3	0.43	0.52	1.6	0.20	0.28	28	n.a.
TUV, 300 DU, 30° N, 21 Dec., s.l.	n.a.	33	0.6	n.a.	100	1.6	0.09	0.14	0.73	0.035	0.08	16	n.a.
TUV, 350 DU, 60° N, 21 Jun., s.l.	n.a.	46	1.0	n.a.	141	2.7	0.17	0.24	1.1	0.07	0.13	19	n.a.
TUV, 350 DU, 60° N, 21 Dec., s.l.	n.a.	3.6	0.01	n.a.	10.9	0.02	0.0005	0.003	0.07	0.001	0.003	1.7	n.a.
Jokioinen, 21 May, 7:30, CMF=1.0	198	23	0.27	907	71	0.71	0.025	0.055	0.48	0.011	0.035	9.9	0.78
Jokioinen, 21 May, 9:30, CMF=1.0	295	37	0.68	1350	113	1.8	0.08	0.14	0.81	0.03	0.08	15.5	1.30
Jokioinen, 21 May, 11:30, CMF=1.0	343	44	0.9	1570	135	2.5	0.13	0.19	1.0	0.05	0.11	18.2	1.56
Helsinki, 8 June 2011, no filter	350	44	1.2	1604	134	3.2	0.25	0.32	1.1	0.12	0.17	23.7	1.98
Helsinki, 8 June 2011, + polythene	324	38	1.0	1487	116	2.7	0.21	0.26	0.93	0.10	0.15	23.0	1.80
Helsinki, 8 June 2011, + polyester	316	33	0.06	1448	102	0.15	0.008	0.028	0.65	0.006	0.03	1.48	0.10
Helsinki, 8 June 2011, + # 226	316	3.2	0.04	1453	10.0	0.11	0.006	0.009	0.04	0.002	0.005	11.0	0.076
UVB313 + diacetate (40 cm)	n.a.	1.6	1.2	n.a.	4.6	3.0	0.53	0.54	0.51	0.47	0.31	658	n.a.

fibre at the spectrometer end[10][11]. Doing so invalidates the calibration because the alignment of the fibre with respect to the entrance slit may be different after the fibre is reattached.

10. When measuring, take into account the field of view of the entrance optics of your instrument (e.g. one hemisphere for a cosine corrected irradiance sensor) and make sure that yourself and any other nearby objects do not disturb the radiation 'seen' by the instrument.

11. Take into account that spot measurements of UV in sunlight under different filters describe only one point in time. Continuous measurements or modelling based on continuous ground-based measurements are needed to fully describe the treatments applied.

12. Outdoor UV irradiance is affected by cloudiness, so measurements where a single instrument is moved to take sequential measurements under the different treatments should be avoided unless the sky is perfectly clear. In addition, parallel measurement of PAR or global radiation in the open is recommended so as to be able to detect any variation due to clouds.

13. UV exposure values (also called doses) derived from satellite-based measurements are subject to relatively large errors and bias, so it is better to avoid their use, specially when daily or weekly values are needed. Much of the uncertainty derives from the sparse nature of the satellite data (e.g. one or fewer fly overs per day).

14. Routinely check your instruments. Frequently check that the readings are very close to zero when the sensor is in darkness. Do sanity checks on your data against values expected (e.g. using models or published data). For example if when measuring a sunlight spectrum you get spectral irradiance values different from zero[12] at wavelengths shorter than 290 nm you can be sure that there is a problem. There may be too much stray light, or a bad correction for the dark signal, or simply the spectrometer is not good enough for the job.

15. When measuring sunlight or lamp spectra for calculating effective UV doses you need a spectroradiometer with a spectral resolution of at least 1 nm.

Furthermore to reduce noise use only averaging of repeated scans rather than 'Boxcar smoothing'. Boxcar smoothing reduces the spectral resolution by doing a moving average across wavelengths.

3.16 Further reading

3.16.1 UV climatology and modelling

`http://cprm.acd.ucar.edu/Models/TUV/`
`Interactive_TUV/`
`http://zardoz.nilu.no/~olaeng/fastrt/`
`http://jwocky.gsfc.nasa.gov/teacher/`
`ozone_overhead.html`

3.16.2 Instrumentation and UV measurement validation

The report *A Guide to Spectroradiometry: Instruments & Applications for the Ultraviolet* (Bentham, 1997) describes scanning spectroradiometers. It is available at `http://www.bentham.co.uk/pdf/UVGuide.pdf`. The report *Instruments to Measure Solar Ultraviolet Radiation - Part 4: Array Spectroradiometers* (Seckmeyer, Bais, Bernhard, Blumthaler, Drüke et al., 2010) gives guidelines for the use of array spectrometers for measuring UV-B radiation. This report is available at `http://www.wmo.int/pages/prog/arep/gaw/documents/GAW191_TD_No_1538_web.pdf`. The report *Instruments to Measure Solar Ultraviolet Radiation - Part 3: Multi-channel filter instruments* (Seckmeyer, Bais, Bernhard, Blumthaler, Johnsen et al., 2010) is available at `http://www.wmo.int/pages/prog/arep/gaw/documents/GAW190_TD_No_1537_web.pdf`. The report *Instruments to Measure Solar Ultraviolet Radiation Part 2: Broadband Instruments Measuring Erythemally Weighted Solar Irradiance* (Seckmeyer et al., 2005) is available at `ftp://ftp.wmo.int/Documents/PublicWeb/arep/gaw/final_gaw164_bookmarks_17jul.pdf`, while the first report of the series, titled *Instruments to Measure Solar Ultraviolet Radiation - Part 1: Spectral Instruments* (Seckmeyer et al., 2001) is only available in printed form.

---

[10]SMA connectors were originally designed for attaching light fibres used in digital communications, and do not guarantee repeatability of the fibre's alignment. Other types of connectors (e.g. FC) that provide better repeatability are available as an additional cost option from some spectrometer manufacturers.

[11]Detaching the fibre from the diffuser does not invalidate the calibration as small changes in alignment will not change the amount of radiation entering the fibre.

[12]You may be thinking how close to zero is close enough. For reliably measuring effective UV doses in sunlight using a spectroradiometer one needs at least four or five orders of magnitude between the highest peak in the visible and the noise signal.

3.16.3 Books

The book *Radiation Measurement in Photobiology* edited by Diffey (1989) includes information on detectors and methods not described in this handbook, or that are described here in less detail.

3.17 Appendix: Formulations for action spectra used as BSWFs

The calculation of biologically effective UV doses requires weighting the spectral energy irradiance (E) or spectral photon irradiance (Q) at each wavelength with a weighting value that is typically generated from a mathematical formulation fitted to describe the weighting function or action spectrum of the biological process—i.e. a biological spectral weighting function (BSWF). Data obtained from a spectroradiometer are usually provided as spectral energy irradiance (units: $W\,m^{-2}\,nm^{-1}$) and care should be taken to use the appropriate formulation that uses either energy effectiveness (s) or quantum effectiveness (s^p) values as appropriate.

The usual approach is to base effective UV doses on energy irradiance data, using BSWFs that provide values for relative energy effectiveness at each wavelength. However, BSWFs are often defined using quantum effectiveness for application to photon irradiance values. Consequently, conversion of energy irradiance values measured during experiments to photon irradiance is necessary before applying weighting functions originally formulated using quantum effectiveness. Conversion of energy irradiance values to photon irradiance can be achieved using equation 3.1 on page 72. After conversion, the calculated effective doses will be on a different scale than if they are calculated based on energy irradiance values. Note that you may find examples in the literature where formulations based on quantum effectiveness (s^p) have been inappropriately applied to spectral energy exposure (unit: $J\,m^{-2}\,nm^{-1}$) or spectral energy irradiance (unit: $W\,m^{-2}\,nm^{-1}$).

The general plant action spectrum of Caldwell (Code GEN in Table 1.4 on page 25) was originally published simply as a graphical figure of quantum effectiveness against wavelength (Caldwell, 1971) and subsequently two publications have fitted different mathematical formulations to describe its use as a weighting function. Green et al. (1974) fitted a function that decreases to zero at 313 nm whereas an alternative mathematical fit provided by Thimijan et al. (1978) continues to weight irradiance values up to 345 nm in the UV-A region. Both of these functions are shown on Figure 1.19 on page 25 as GEN(G) and GEN(T) respectively. It is important to specify which mathematical function has been used when describing the calculation of effective UV doses in experimental methods.

The general plant action spectrum of Caldwell fitted with the mathematical function of Green et al. (1974) is given by (Source: Björn and Teramura, 1993):

$$s^p_{GEN(G)}(\lambda) = \begin{cases} 2.618 \cdot \left[1 - \left(\frac{\lambda}{313.3}\right)^2\right] \cdot e^{-\frac{\lambda-300}{31.08}} & \text{if } \lambda \leq 313.3 \text{ nm} \\ 0 & \text{if } \lambda > 313.3 \text{ nm} \end{cases} \tag{3.7}$$

and when fitted with the mathematical function of Thimijan et al. (1978) is given by (Source: Björn and Teramura, 1993):

$$s^p_{GEN(T)}(\lambda) = \begin{cases} e^{-\left(\frac{265-\lambda}{21}\right)^2} & \text{if } \lambda \leq 345 \text{ nm} \\ 0 & \text{if } \lambda > 345 \text{ nm} \end{cases} \tag{3.8}$$

The DNA damage formulation of Green and Miller (1975) is given by:

$$s^p_{DNA(GM)}(\lambda) = e^{13.82 \cdot \left[\left(1 + e^{\frac{\lambda-310}{9}}\right)^{-1} - 1\right]} \tag{3.9}$$

The doses calculated with this formulation differ significantly from doses calculated using tabulated values derived from the figure in Setlow (1974). For example the model TUV and the data from the NSF UV monitoring network use the tabulated values rather than the formulation by Green and Miller (1975).

However, most BSWFs are conventionally used with a value of one at 300 nm. This may be achieved by simple calculation adjustments within a spreadsheet (by dividing each wavelength effectiveness by the effectiveness value at 300 nm) or by altering the mathematical formula directly. Thus, the mathematical formulation of Green et al. (1974) for GEN(G) requires equation 3.7 to be multiplied by 4.596 to normalize it to a value of 1 at 300nm and similarly GEN(T) requires equation 3.8 to be multiplied by 16.083 and DNA(GM) requires equation 3.9 to be multiplied by 30.675.

The formulation of the weighting function published by Flint and Caldwell (2003) for plant growth, shown as PG in Figure 1.19 on page 25, was for quantum effectiveness and already provides normalization to 1 at 300 nm and is given by:

$$s_{PG}^p(\lambda) = \begin{cases} \exp\left(4.688272 \cdot \exp\left(-\exp\left(0.1703411 \cdot \frac{\lambda - 307.867}{1.15}\right)\right) + \left(\frac{390-\lambda}{121.7557} - 4.183832\right)\right) & \text{if } \lambda \leq 390 \text{ nm} \\ 0 & \text{if } \lambda > 390 \text{ nm} \end{cases} \quad (3.10)$$

Some weighting functions are already defined using energy effectiveness values for use with energy irradiance data, one example being the CIE erythemal action spectrum (McKinlay and Diffey, 1987; A. R. Webb et al., 2011). The standard was revised in 1998 and the updated version should be used instead of the original one from 1987 (A. R. Webb et al., 2011). It is important to check that published mathematical formulae have been appropriately normalized to one at 300 nm. However, whereas most BSWFs are conventionally set to one at 300 nm, the CIE98 erythemal weighting function has defined values at specific wavelengths (A. R. Webb et al., 2011) and is given by:

$$s_{CIE}(\lambda) = \begin{cases} 1 & \text{if } 250 \leq \lambda \leq 298 \text{ nm} \\ 10^{0.094(298-\lambda)} & \text{if } 298 \leq \lambda \leq 328 \text{ nm} \\ 10^{0.015(140-\lambda)} & \text{if } 328 \leq \lambda \leq 400 \text{ nm} \\ 0 & \text{if } \lambda > 400 \end{cases} \quad (3.11)$$

Weighting functions defined using quantum effectiveness ($s^p(\lambda)$), and normalized to one at 300 nm, can be converted to relative energy effectiveness ($s(\lambda)$) simply by multiplication, wavelength by wavelength, of the quantum effectiveness by by the respective wavelength in nm and dividing by 300 (the chosen normalization wavelength in nm). The value 300 should be changed when using other wavelengths for normalization. This is based on equation 3.1 giving the energy in one photon, but as Planck's constant (h) and the speed of light in vacuum (c) are constant divisors, and we are expressing the effectiveness in relative units, they can be left out of the calculation.

It is always essential to specify clearly the normalization wavelength, the mathematical formulation and the BSWF used when describing the calculation of effective UV doses.

3.18 Appendix: Calculation of effective doses with Excel

Let us start with a text file generated by a spectroradiometer and its connected computer. With the instruments that I have used this text file consists of one column, containing either spectral (energy) irradiance or spectral photon irradiance (some instruments, especially for underwater use, may have fluence rate instead of irradiance). At the start, above the columns, is a heading containing supplementary information. Other instruments may have two columns, the first one containing wavelength values, the other one spectral irradiance values.

For the Optronics instruments that I have used, the start of a data file looks like the listing in Figure 3.32. The heading here contains "a", which in this case is the name of the file, the kind of data that is recorded (including the unit), the date, the starting wavelength in nm, the end wavelength (not shown here) in nm, and the step interval in nm. This file can be loaded into an Excel file going to `Data>Get external data>Load text file`. When you have done this, insert a new column to the left of the one containing the measured data. In this new column you should fill in the wavelength values, which can be done in the following way without typing every value:

Type the first two values, i.e. in this case first "250" in the cell left of the one containing the first radiation value 1.439562E-010. Since the step interval is 1.00 nm, the next value is 251. Type this in the cell to the left of the value 1.797497E-010. Select the two cells that you have just filled in. Put the cursor at the lower left corner of the cell containing "251" and see it change appearance, push the (left) mouse button and drag down to the cell left of the last radiation value. The column will then fill up with the appropriate wavelength values.

You have now filled columns A and B with the appropriate values. Next you should generate a column with values of the weighting function. Suppose you wish to use the weighting function of Flint and Caldwell (2003). The formula given for this in the publication is Biologically effective UV= exp[4.688272 ∗ exp(− exp(0.1703411 ∗ $(x-307.867)/1.15)) + ((390-x)/121.7557 - 4.183832)]$ in which x stands for wavelength in nm.

Select cells in column C over the rows corresponding to values in columns A and B. Go to "Tools" on top of the file, push the mouse button and go down and select "Calculator". Type the formula above, with the exceptions that you use ordinary parentheses instead of square brackets,

```
"a","Irradiance [W/(cm^2 nm)]",971217,250.00,400.00,1.00
1.439562E-010
1.797497E-010
6.126532E-010
4.516362E-009
8.027722E-009
1.081851E-009
2.862175E-010
1.728253E-010
1.790998E-010
1.953554E-010
2.570309E-010
...
```

Figure 3.32: Top of a data file from an Optronics spectroradiometer.

"A:A" instead of "x" (do not type the quotation marks), and comma instead of dot if you use the comma system. When this is done, press OK and save the result.

You now have the values of the weighting function in column C. If you double-click on the first value in column C you should get the equation entered. The top of your file should then look like Figure 3.33. If you insert a new column C to the left of your old column C and copy the wavelength values into it you can plot your weighting function as a check. If you plot it on a logarithmic abscissa, it should look something like Figure 3.34. This can be compared to the plot in the original publication by Flint and Caldwell (2003). Once you have generated a weighting function that you wish to use several times, you need of course not calculate it like this every time. You can simply transfer the column with it to a new sheet or a new Excel file. Since, as newly generated, it depends on another column, you need to use the command "Paste special" and choose "values".

Now you are ready to do the weighting itself. Select cells in column E in rows corresponding to the values in the other columns. Select the Calculator, and use it to multiply values in column B with the weighting function. Select an empty cell and go a final time to the Calculator. Select "Sum" and then the values you have just generated in column E. The sum you get in the cell you selected is the weighted radiation value. The number is, in this case, 0.000334759 and the unit is W cm^{-2}, which can also be expressed as 3.35 W m^{-2}. The bottom of the file should look like Figure 3.35. I have written Σ in one cell to remember that the value to the right of it is the sum of the values above it.

Note that if you have another step size than 1 nm, you must multiply the sum by that step size. It is recommended that you do not use step sizes greater than 1 nm for UV-B spectra, since both the spectra themselves and the weighting functions are so steep.

Martyn Caldwell's traditional Generalized Plant Action Spectrum is easier to handle, and you should be able now to do a similar exercise with it yourself. A formula for

this weighting function has been published by Green et al. (1974): Weighting function = $2.618 \cdot [1 - (\lambda/313.3)2] \cdot \exp[-(\lambda - 300)/31.08]$, where λ stands for wavelength in nm. We abbreviate the name of this action spectrum as GEN(G) elsewhere in this text. See section 3.17 on page 111 for the equations for other commonly used BSWFs.

3.19 Appendix: Calculation of effective doses with R

3.19.1 Introduction

If you use the R system for statistics (see, `http://www.r-project.org/`), or if you need an implementation with fewer restrictions, you may want to use R instead of Excel to calculate doses and action spectra. R is based on a real programming language called S and allows much flexibility. We have developed a package to facilitate these calculations. The package is called 'UVcalc' and will be soon submitted to CRAN (Comprehensive R archive network) and will be available also from this handbook's web page. In addition to functions for calculating weighted and unweighted UV doses, and PAR irradiance and PPFD from energy irradiance spectra, also functions for calculating photon ratios are provided.

3.19.2 Calculating doses

Currently functions for five BSWFs are implemented and are listed in Table 3.9. The functions take two arguments one vector giving the wavelengths and another vector giving the values of spectral energy irradiance at these wavelengths. The spectral irradiance or spectral exposure values would come either from measurements with a spectroradiometer or from model simulations. All functions accept a wavelengths vector with variable and arbitrary step sizes, with the condition that the wavelengths are sorted in strictly increasing order, something which is especially convenient when dealing with data from array

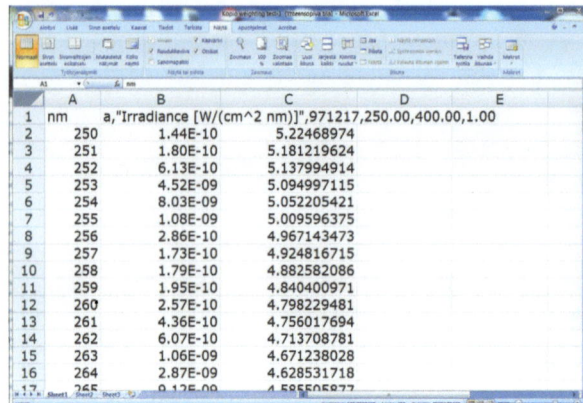

Figure 3.33: Screen capture from a Excel worksheet onto which spectral data have been imported.

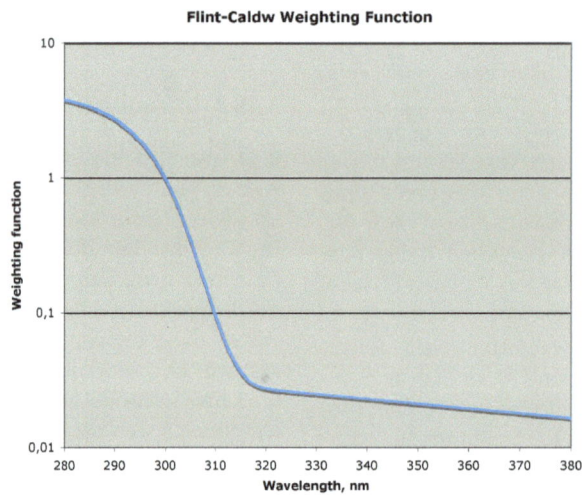

Figure 3.34: Screen capture from a Excel plot of the plant growth action spectrum (PG).

148	396	2,08E-07	396	0,014507188	3,01988E-09
149	397	2,10E-07	397	0,014388526	3,01462E-09
150	398	2,20E-07	398	0,014270835	3,14568E-09
151	399	2,19E-07	399	0,014154106	3,09404E-09
152	400	2,26E-07	400	0,014038332	3,17621E-09
153					
154					
155				Σ	0,000334759

Figure 3.35: Screen capture from the bottom of an Excel worksheet on which doses have been calculated.

Table 3.9: Functions in R package UV.calc for calculation of unweighted and effective irradiances or doses, depending on whether they are used with an irradiance spectrum, or with daily total spectral exposure. The functions for effective (or weighted) irradiances available in this package use different action spectra as BSWF, normalized to unity at different wavelengths (λ). w = vector of wavelengths (nm); i = vector of spectral energy irradiances.

Action spectrum	Formulation	Dose or irrad.	Norm. λ (nm)
Gen. plant action	Green	GEN.G.dose(w,i)	300
Gen. plant action	Thimijan	GEN.T.dose(w,i)	300
Plant growth	Flint & Caldwell	PG.dose(w,i)	300
Erythemal	CIE98	CIE.dose(w,i)	298
'Naked' DNA	TUV, from Setlow	DNA.N.dose(w,i)	300
'Naked' DNA	Green & Miller	DNA.GM.dose(w,i)	300
'Plant' DNA	Musil, from Quaite	DNA.P.dose(w,i)	300
UV-A (energy)	n.a.	UVA.e.dose(w,i)	n.a.
UV-A (photon)	n.a.	UVA.q.dose(w,i)	n.a.
UV-B (energy)	n.a.	UVB.e.dose(w,i)	n.a.
UV-B (photon)	n.a.	UVB.q.dose(w,i)	n.a.
PAR (energy)	n.a.	PAR.e.dose(w,i)	n.a.
PAR (photon)	n.a.	PAR.q.dose(w,i)	n.a.

spectrometers[13].

The functions are made available by installing the package UVcalc (once) and loading it from the library when needed. To load the package into the workspace use library(UVcalc). Then load your spectral data into R using read.table() or read.csv().

A file from a Macam spectroradiometer starts:

```
Wavelength (nm) ,W/m2
270,0
271,0
272,1.17E-04
273,2.42E-04
274,4.55E-04
275,8.94E-04
276,0.00161
277,0.00263
278,0.00412
279,0.00621
280,0.00904
281,0.01697
282,0.02069
283,0.02663
284,0.03314
285,0.04075
286,0.04895
287,0.05817
288,0.0679
```

For a file like this one, use the code below but replacing "name" with the name and path to the data file. On Windows systems you need to scape backslashes in file paths like this: '\\'.

```
my.data <- \read.csv(filename="name", skip=1,
    col.names=c("wavelength", "irrad"))
attach(my.data)
GEN.G.dose(wavelength,irrad)
PG.dose(wavelength,irrad)
detach(my.data)
```

If our spectral irradiance data is in $W\,m^{-2}\,nm^{-1}$, and the wavelength in nm, as in the case of the Macam spectroradiometer, the functions will return the effective irradiance in $W\,m^{-2}$.

If, for example, the spectral irradiance output by our model or spectroradiometer is in $mW\,m^{-2}\,nm^{-1}$, and the wavelengths are in Ångstrom then to obtain the effective irradiance in $W\,m^{-2}$ we will need to convert the units.

```
...
GEN.G.dose(wavelength/10, irrad/1000)
PG.dose(wavelength/10, irrad/1000)
...
```

In this example, we take advantage of the behaviour of the S language: an operation between a scalar and vector, is equivalent to applying this operation to each member of the vector. Consequently, in the code above, each value from the vector of wavelengths is divided by 10, and each value in the vector of spectral irradiances is divided by 1000.

If the spectral irradiance is in $mW\,cm^{-2}\,nm^{-1}$ then values should be multiplied by 10 to convert them to $W\,m^{-2}\,nm^{-1}$.

It is very important to make sure that the wavelengths are in nanometers as this is what the functions expect. If the wavelengths are in the wrong units, the BSWF will be wrongly calculated, and the returned value for effective irradiance will be wrong.

If we use as input to the functions instead of spectral irradiances, time-integrated spectral irradiances in $kJ\,m^{-2}\,d^{-1}\,nm^{-1}$, then the functions will return the ef-

---

[13]It is always best to use in calculations the data at the original wavelength of each pixel of the array of the spectrometer, rather than to produce a spectrum with 'nice' wavelength steps by interpolation.

fective exposure, or 'dose' in $kJ\,m^{-2}\,d^{-1}$. Such time-integrated values are more frequently available as the output of models, or by integrating observed sequential values of spectral irradiance.

In addition to the functions for calculating biologically effective doses and effective irradiances, we provide functions for calculating unweighted doses or irradiances of PAR (`PPFD()` or `PAR.q.dose()`, and `PAR.e.dose()`), UV-A (`UVA.q.dose()` and `UVA.e.dose()`) and UV-B (`UVB.q.dose()` and `UVB.e.dose()`), where 'e' variants return energy doses, and 'q' variants return quantum or photon doses. All functions expect as input radiation spectra in energy units, and wavelengths in nm.

3.19.3 Calculating an action spectrum at given wavelengths

The functions available for calculating action spectra take as argument a vector of wavelengths, and return a vector of effectiveness (either quantum=photon or energy based) normalized to unity effectiveness at a wavelength of 300 nm except when indicated. These functions are listed in Table 3.10, and an example of their use follows. In these examples we generate the wavelengths vectors in R, but they can be also read from a file.

```
# at 1 nm intervals
wavelengths1 <- 285:400
action.spectrum1 <- GEN.T.e(wavelengths1)
# at 0.1 nm intervals
wavelengths01 <- seq(from = 285, to = 400, by = 0.1)
action.spectrum01 <- GEN.T.e(wavelengths01)
```

All functions accept a wavelengths vector with variable and arbitrary step sizes, with the condition that the wavelengths are sorted in strictly increasing order.

3.19.4 Calculating photon ratios

Functions are also provided for calculating photon ratios between different pairs of wavebands. These functions are listed in Table 3.11. We follow the most frequently used wavelength ranges for the different colours, but also provide some generic functions that can be used when other limits are needed. Continuing with the example above in which my.data was read and attached we calculate UV-B : PAR photon ratio

`UVB.PAR.ratio(wavelengths, irrad)`

The spectral energy irradiance, can be in any energy based unit such as $W\,m^{-2}\,nm^{-1}$, or $mW\,cm^{-2}\,nm^{-1}$, as the multipliers cancel out when calculating the ratio. However, wavelengths must be always expressed in nanometers. Please, be aware that these functions will return erroneous values if used with spectra expressed as spectral photon irradiance, even though the returned values are photon ratios.

Please, be aware that following common practice in the literature, the wavelength range used for red light is different for the different photon ratios.

3.19.5 Documentation

The package includes manual pages for the different functions, and an overview and a list of references for the original sources used. The definition of the functions can bee seen once the package is loaded by entering the name of the function without parameters or parentheses.

There are many resources on the R-System for Statistics and statistics in general available on-line. The most useful ones are The R Wiki (`http://rwiki.sciviews.org/`), and the documentation included in all R installations and accompanying each one of the different packages.

Some classical books on R are those by Dalgaard (2002), Crawley (2005, 2007), and Venables and Ripley (2002). The book by Venables and Ripley, 2000 discusses programming in S, the language used in R to build new functions and packages.

3.20 Appendix: Suppliers of instruments

In this section we provide names and web addresses to some suppliers of instruments. This is certainly an incomplete list and exclusion reflects only our ignorance.

UV measurement:

`http://www.astranetsystems.com/`
`http://www.avantes.com/`
`http://www.bentham.co.uk/`
`http://www.biosense.de/` ('*Viospore*' dosimeters.)
`http://www.biospherical.com/`
`http://www.deltaohm.com/`
`http://www.delta-t.co.uk/`
`http://www.gigahertz-optik.de/`
`http://www.goochandhousego.com/` ('*Optronics*' spectroradiometers)
`http://www.ictinternational.com.au/` (ELDONET terrestrial dosimeters)
`http://www.intl-lighttech.com/`
`http://www.irradian.co.uk/` ('*Macam*' instruments under new name)
`http://www.kippzonen.com/`
`http://www.oceanoptics.eu/`
`http://www.roithner-laser.com/` (Photodiodes)
`http://www.scitec.uk.com/`
`http://www.sglux.com/` (SiC photodiodes, and instruments)
`http://www.skyeinstruments.com/`

Table 3.10: Functions in R package UVcalc for calculation of action spectra. w = vector of wavelengths (nm).

Action spectrum	Formulation	Energy	Quantum	Norm. λ (nm)
Gen. plant action	Green	GEN.G.e(w)	GEN.G.q(w)	300
Gen. plant action	Thimijan	GEN.T.e(w)	GEN.T.q(w)	300
Plant growth	Flint & Caldwell	PG.e(w)	PG.q(w)	300
Erythemal	CIE98	CIE.e(w)	CIE.q(w)	298
Erythemal	CIE98	CIE.e300(w)	CIE.q300(w)	300
'Naked' DNA	TUV, from Setlow	DNA.N.e(w)	DNA.N.q(w)	300
'Naked' DNA	Green & Miller	DNA.GM.e(w)	DNA.GM.q(w)	300
'Plant' DNA	Musil, from Quaite	DNA.P.e(w)	DNA.P.q(w)	300
'Plant' DNA	Musil, from Quaite	DNA.P.e290(w)	DNA.P.q290(w)	290

Table 3.11: Functions in R package UVcalc for calculation of photon ratios from spectra in spectral energy units. w = vector of wavelengths (nm); i = vector of spectral energy irradiances.

Ratio	R function	wavelength ranges (nm)
UV-B:UV-A	UVB.UVA.ratio(w,i)	280–315, 315–400
UV-B:PAR	UVB.PAR.ratio(w,i)	280–315, 400–700
UV-A:PAR	UVA.PAR.ratio(w,i)	315–400, 400–700
Red:Far-red	R.FR.ratio(w,i)	650–670, 720–740
Blue:Green	B.G.ratio(w,i)	420–490, 500–570
Blue:Red	B.R.ratio(w,i)	420–490, 620–680

http://www.solarlight.com/
http://www.spectralproducts.com/
http://www.stellarnet.us/
http://trios.de/ (RAMSES radiometers)

http://www.yesinc.com/ (Yankee Environmental Systems)

4 Plant growing conditions

Eva Rosenqvist, Félix López Figueroa, Iván Gómez, Pedro J. Aphalo

4.1 Introduction

By applying the most constant growth conditions that we can in an experiment, we try to minimise uncontrolled variation in our measured plant data, even though this is not always achieved even with identical climate chamber protocols (Massonnet et al., 2010). However, by growing plants in stable controlled environments we will also induce differences in their response to any given stress, since natural variation in the outdoor climate contributes to plants' ability to cope with stress. Plants grown in climate chambers are "softer" than plants grown in greenhouses, that are themselves "softer" than plants grown outdoors. This softness can be expressed as weaker stems, physically softer leaves with less developed cuticles, and a lessened ability to regulate their stomata.

Any protected cultivation will invariably differ from natural outdoor conditions, and the relative importance of these differences in aspects of the climate will vary between controlled and natural environments. The most pronounced difference is the constancy of a created climate under protected cultivation—and the lack of wind and UV-B radiation! In experiments in greenhouses during the winter and in growth chambers, the amount of radiation received by plants will be far less than they get in summer outside when most plants are active. This is unrealistic, unless dealing with obligate shade species.

4.2 Greenhouses

Greenhouses come in many shapes and materials throughout the world. For plant production the cladding material can be either glass or plastic. Normal glass is opaque to UV-B and only expensive quartz glass allows UV-B to penetrate. The latter is used in only a few places where near-ambient UV-B radiation is considered important for ecophysiological research, but the vast majority of glasshouses provide no natural UV-B inside. The use of plastic opens several opportunities for choosing materials that are transparent to UV-B. Plastics are available as UV-B opaque and UV-transparent films that transmit <1% or >70% of the radiation < 400 nm, respectively. They can be clear (transmitting 87–90% of PAR) or milky to diffuse the light (transmitting 85% of PAR). UV-B transparent films are yet primarily used for increasing the vegetable quality in horticultural production in warmer climates.

This chapter will by no means give complete instructions on how to manage the climate in a greenhouse, but we have gathered some information about possible technical solutions to various climate problems and some tips and tricks from personal experience. Greenhouses are available that provide different technical solutions for controlling the climate. In this section we will focus on glass greenhouses with fully automatic climate control regulated by a climate computer. The topic of how to apply UV-B from lamps has already been covered in section 2.2 on page 36.

4.2.1 Temperature

It is technically easier to heat than to cool a greenhouse and it is therefore easier to maintain a stable temperature during cold times of the year than during warm weather. As long as the boiler and heating system are correctly set-up and able to cope with the maximum expected temperature difference between outside and inside, the temperature should remain close to the point set by the thermostat. The amount of variation around the set point will depend on the sensitivity of the on/off signal to the boiler from the thermostat that regulates the heating system. Where winters are particularly cold, some greenhouses require insulation screens to enable them to maintain e.g. 20°C air temperature.

Greenhouses with conventional air conditioning are not common due to their high running costs, so cooling is normally done by passive ventilation, misting systems or fans with cooling pads. Passive cooling involves opening vents and/or using shade screens to decrease heat load from the sun. The temperature outside will naturally have a huge impact on the degree of cooling achieved by ventilation and, if the shade screens are used for cooling, it is important to keep a 10% gap between them to allow air circulation out of the greenhouse.

In a closed greenhouse in sunshine strong temperature gradients can be created. When the temperature at the floor level is ca. 25°C, it may be 30°C just 35 cm above the floor and >50°C just under the ridge of the roof. A few horizontal fans with moderate rotation in the ceiling can mitigate the problem. Regardless if fans are used or not it is important that the climate sensors connected to the climate computer are placed close to the plants since the plants respond to the microclimate in the canopy, not the macroclimate somewhere else in the greenhouse. It is also crucial that the temperature sensor is shielded from direct heat radiation from the sun or lamps to ensure that the true air temperature is measured.

Water-based cooling systems use the principle of evaporative cooling. As water evaporates, some sensible heat from the warm air is transferred to the water as latent heat during its vaporisation. These cooling systems work best in areas with low to moderate air humidity. Misting systems can decrease the air temperature by some °C in moderate humidity and can be constructed with or without fans. It is important that they create droplets that are small enough to evaporate completely before they have fallen down onto the plants. "Rain" under misting systems will drastically increase the risk of fungal disease on plants.

Fans with cooling pads regulate temperature using large industrial fans mounted in series with pads where water trickles down an enlarged surface (e.g. corrugated fibreboard, wood shavings or wheat straw), which enhance the area for evaporation. The change in latent heat in the evaporated water leads to a corresponding decrease in air temperature.

4.2.2 Light

Greenhouse conditions should mimic the outdoor conditions of the open landscape as much as possible. The light environment in a greenhouse is strongly dependent on how much shade is created by its structure. Modern greenhouses are constructed to keep shading to a minimum.

The amount of radiation received by the plants can be increased by using lamps and decreased by using shade screens but regardless of technical installations

it will be difficult to create summer light conditions in a greenhouse during the winter in latitudes covering most of Europe. During the summer in Denmark daily integrals of PAR vary in the range of ca. 10–30 $mol\,m^{-2}\,d^{-1}$ between rainy and sunny days inside a greenhouse without shade screens, while the values in midwinter are 0.5–5 $mol\,m^{-2}\,d^{-1}$ with some use of supplemental light until 22:00 h in the evening (Figure 4.1).

Modern greenhouses have two or three screens for different purposes (shade, insulation, blackout) but in most greenhouses one shade screen can perform a dual function in insulation as well as shading radiation. The shade screen may also differentially remove part of the radiation, allowing the user to decide what fraction of the radiation load to remove up to the maximum shading with the screen fully closed. In many cases, the principal role of shade screens is to decrease the heat load from the sun, preventing the greenhouse from getting too hot, or as insulation screen during the cold season, but we should bear in mind that the light environment created by a medium-to-dense screen is comparable to that of an overcast day i.e. the PAR is decreased by 10–90%, depending on screen material.

For plants whose flowering is controlled by day length, blackout screens can be used to give short day treatments and incandescent bulbs (low red:far-red photon ratio) for night interruption to achieve long day treatments[1]. Either a single pulse, or multiple pulses—e.g. 5 min per hour—can be used in practice. In this way flowering can either be induced or plants kept in a vegetative state for both short- or long-day plants. For the treatment to work, the blackout screens must provide completely black conditions without gaps, since phytochrome responds to extremely small amounts of light, and even small cracks between screens would be enough to interrupt the night length signal.

During the winter season, supplemental light from lamps is needed for most species grown in greenhouses at high latitudes. Most greenhouses in these regions are supplied with high-pressure sodium (HPS) lamps, which give a spectrum from blue to red but predominantly emit orange light (Figure 4.2). For photosynthesis, this spectrum is not optimal but when it is combined with natural daylight most plants grow without any problems. In research greenhouses, high-pressure lamps based on elements other than sodium are also used to give more "white" light. However, they are not used for plant production because of their lower energy efficiency than HPS lamps.

Light-emitting diodes (LEDs) are currently being developed to provide greenhouse lighting. They come in

---

[1]The length of the night is the signal perceived by plants, and this is the reason why an interruption of the night is equivalent to long day conditions.

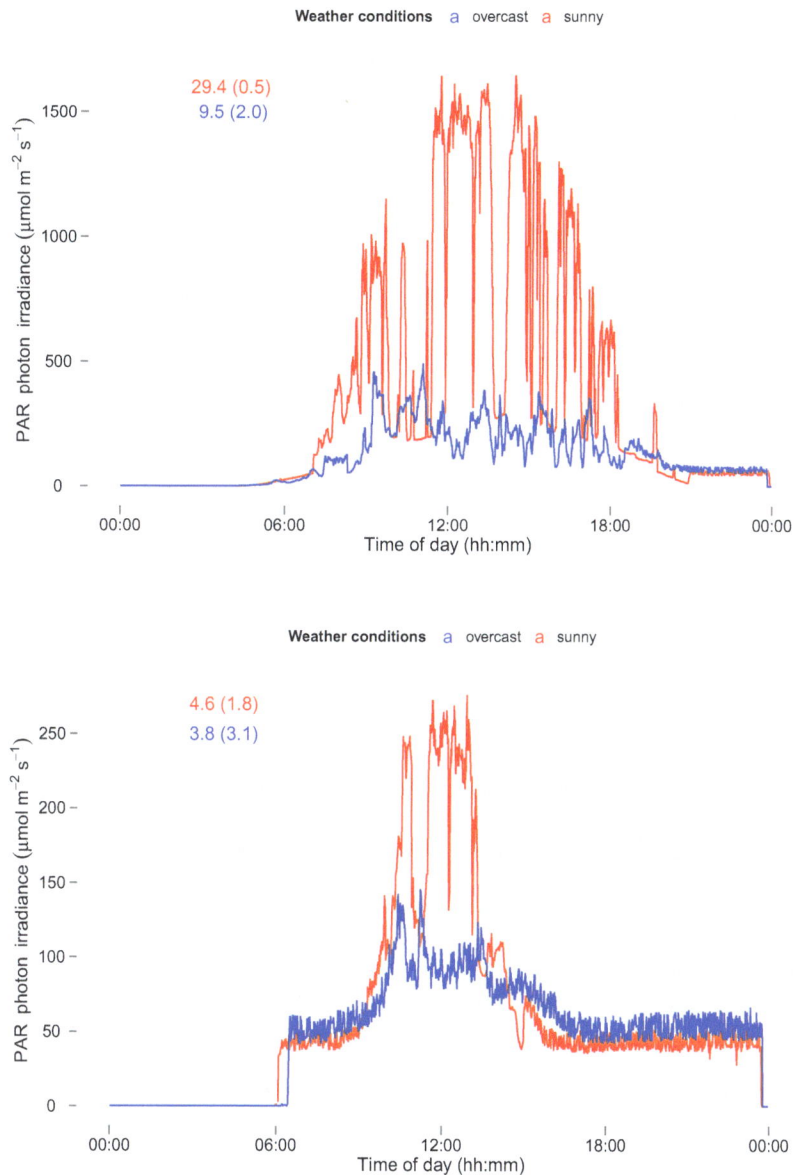

Figure 4.1: PAR photon irradiance in a greenhouse in Denmark, on sunny days with broken clouds (red) and overcast (blue) days, during summer (top) and winter (bottom). Whole-day integrated PAR photon exposures ($mol\,m^{-2}\,d^{-1}$) are indicated on the figures, with the contribution from lamps in brackets. The lamps were controlled by the climate computer to turn on when the light level decreased below the set point and to extend the day. Note that the change in PPFD can be up to $\approx 1500\,\mu mol\,m^{-2}\,s^{-1}$ from one minute to the next in sunny weather with scattered clouds. In greenhouses also the shade from the structure contribute to the fluctuations in PPFD. Note the different y-scale for summer and winter. The signal noise seen at low PAR is picked up by the sensor (here a LiCor quantum sensor connected to a CR10 Campbell data logger) from the HPS lamps when they are turned on in the greenhouse, i.e. data has to be averaged over several measurements to reveal the true light variation pattern.

Figure 4.2: High pressure sodium lamps in a greenhouse in which sunlight is blocked to simulate natural Finnish winter light conditions. Photograph: © Valoya Oy.

many different colours, and can emit anything from a narrow to a broad range of wavelengths. For use in greenhouses, numerous small companies have developed various LED lamps but up to now (summer 2012) large companies like Philips and Osram have as yet only released LED modules for use in growth chambers and for interlighting rows of vegetable crops (long narrow LED units placed between rows of e.g. tomatoes and cucumbers grown in greenhouses). Top light, or "ordinary" greenhouse LED lamps, will require LEDs that are sufficiently efficient to emit enough radiance while incorporating cooling for their electronics. These two factors have to be included in the "budget" when the energy efficiency is calculated for new LED lamps compared to traditional HPS lamps. Both the energy efficiency and lifetime of LEDs will decrease if their electronics are overheated.

Research on the spectral distribution of LED lamps is focusing on red light (which requires the least energy to produce per photon) and how much additional blue or white light needs to be added for good plant growth. One basic problem with this approach is that the human eye is most sensitive to green and yellow and less so to blue and red light. Since the colour of leaves is based on reflected wavelengths, leaves that are only illuminated by blue and red light, which is efficiently absorbed by chlorophyll, will appear as dark objects. It may distort our judgments of plant health and nutritional status when making comparisons, since it is often manifested as yellow chlorosis or reddish coloration of the leaves. One would guess that the technical development of specific greenhouse lamps will lead to red enriched LED lamps with some blue and white (or green) LEDs to improve colour recognition.

Beyond their spectral composition, the main difference between LED and HPS lamps is that the latter produce heat radiated in the direction of the plants while the former do not. This means that the leaf temperature will be higher under HPS lamps than under LED lamps. However, the electronics of the LEDs do become warm. When using only one row of low-efficiency LEDs, the metal mounts for the LED modules often provide enough heat sink to prevent them from over heating (which would otherwise decrease their efficiency), whereas high-efficiency LEDs require cooling. Most early LED illumination for greenhouses was water-cooled and this is one reason that its use in greenhouses has not yet been very widespread. Currently there are some high-power LED lamps available that are passively cooled by convection (Figure 4.3). Some other LED lamps with passive cooling have been designed for research (Figure 4.4) or with active cooling for production (Figure 4.5). The colour of the radiation emitted depends on the LEDs used, and there are currently two different approaches when designing LED lamps: (1) maximising radiation output with respect to electricity consumed, and (2) maximising plant yield and crop quality irrespective of irradiance per unit of electrical power consumed. Because of the present fast technical development of the LED technology we can expect to see numerous new LED products for plant growth on the market in the next couple of years. From the perspective of research into the effects of UV-B in nature, the best PAR sources are those with a spectrum similar to sunlight.

The light distribution in greenhouses can be enhanced by diffusing glass, diffusing plastic films or thin diffusing screens. These create a more even distribution of light and decrease the differences, such as light gradients or patchiness, that are otherwise created by shade from the structure. A resultant side effect of diffusers is enhanced

Figure 4.3: A comparison of the performance of two types of LED lamps with different emission spectra being used for the cultivation of an ornamental species (*Kalanchoe*). Photograph: © Valoya Oy.

Figure 4.4: Philips' adjustable research modules with high efficiency LEDs: (A) white LEDs mounted in climate chamber (for which the research modules are designed) and (B) four modules used in a greenhouse (from left 40% blue in red, 20% blue in red, 100% red and white LEDs). In the climate chamber this setup of modules can be adjusted up to ca 300 $\mu mol\,m^{-2}\,s^{-1}$. Note that the white treatment looks brighter to the eye than the red and blue combinations, despite that all four colours give identical PPFD. Photograph: Eva Rosenqvist.

growth of plants due to better penetration of light into the canopy (Markvart et al., 2010).

Commercially-available climate computers are often connected to photometric (lux) sensors that measure light visible to humans ($\approx$400–700 nm, see Box 1.1 on page 6), whereas global radiation is measured radiometrically ($W\,m^{-2}$, see Table 1.2 on page 5) in the range 400–1100 or 400–3000 nm to provide information for the energy-related control of greenhouses (primarily temperature). The latter is the correct way to measure irradiance linked to climate control; whereas visible light, related to photosynthesis and plant production, should be measured as PAR photon irradiance or photosynthetic photon flux density (PPFD, $\mu mol\,m^{-2}\,s^{-1}$, see Box 1.2 on page 7) or PAR (energy) irradiance ($W\,m^{-2}$). Modern climate com-

puters can also be directly connected to PPFD sensors.

When using supplemental light from various light sources extra care need to be taken on the choice of quantum sensor since most supplemental greenhouse lamps have a dis-continuous spectrum. Some cheaper quantum sensors do not cover the full spectral range 400–700 nm and the sensor may not cover the spectrum of red LED's, which can peak close to a wavelength of 700 nm.

One should keep in mind, that moving an experiment from the field to a greenhouse does not make it independent of the outdoor season, since the UV-B dose should be balanced against the integrated daily PPFD if the experiment aims to mimic outdoor conditions. Daylight varies tremendously between winter and summer. Even

Figure 4.5: Top light LED lamps (Fionia Lighting, DK) tested for plant growth and energy consumption for light and greenhouse heating, in comparison with high pressure sodium lamps in the next greenhouse compartment. The setup gives supplemental PPFD of 120 $\mu mol\, m^{-2}\, s^{-1}$ at a distance of 2.2 m from the bench, which is a level used for high light demanding species in production. Photograph: Carl-Otto Ottosen.

with supplemental light the PPFD of a sunny winter day in Northern Europe will only correspond to an overcast day in the summer (Figure 4.1). The contribution from lamps will be less than what is found in sunlight during the summer, even under long photoperiods (c.f. Figures 4.6 and 4.1).

Except for late successional species that ecologically are adapted to obligate shade, most plants will acclimate to the prevailing light conditions (Bazzaz and Carlson, 1982). This means changing chloroplast architecture and the stoichiometry of the whole photosynthetic apparatus (Anderson et al., 1988) and thus also altering the light response of photosynthesis.

Most experiments on light acclimation only discus the light environment in terms of "high light" or "low light", regardless if it has been created as % of sunlight with permanent shade screens or as a certain PPFD in climate chambers. However, the light environment should be described by day length, maximum PPFD and integrated daily PPFD. Not much is known about which of the parameters of the light environment, actually trigger acclimation of photosynthesis. The three light parameters mentioned above have been investigated in tomato (Van den Boogaard et al., 2001), where the day length affected the chlorophyll concentration while the integrated daily PPFD affected the maximum rate of CO_2 fixation at light saturation, independent of what the maximum PPFD was during the day.

If photosynthesis is being studied in greenhouses in relation to UV-B one should be aware of the effect that the seasonal changes in daylight has on the photosynthetic apparatus and therefore also on the growth of the plants, when comparing different experiments conducted at different times of the year.

4.2.3 Air humidity

Air humidity is difficult to regulate in greenhouses. If few plants are grown in a compartment the humidity tends to be low due to the large areas of potentially cold glass and concrete that act as dehumidifiers through condensation. If the leaf area index (LAI) is high, transpiration will make humidity correspondingly high despite condensation on the greenhouse surfaces.

Misting systems can increase the humidity but it is important that they create mist, not rain. Rain or condensation on the plants will increase the risk of fungal disease in the experimental plants.

The humidity will be higher in highly insulated than in single-glass greenhouses, where some of the humidity will condense on the glass. High air humidity is decreased by opening the vents at the same time as turning on the heating system to increase the exchange of air with the outside. The effect of dehumidifying ventilation will depend on the outdoor weather conditions and on whether dry or moist air is entering the greenhouse.

Using traditional climate control, the air humidity is regulated as relative humidity (RH, %). The driving force of transpiration is the absolute gradient of water vapour from the intercellular spaces inside the leaf to the outside air, i.e. the water vapour pressure difference (Δe, Pa). As the intercellular spaces are microscopic and the cell walls saturated with water it is assumed that the RH in the leaf is 100% at the given leaf temperature. Since

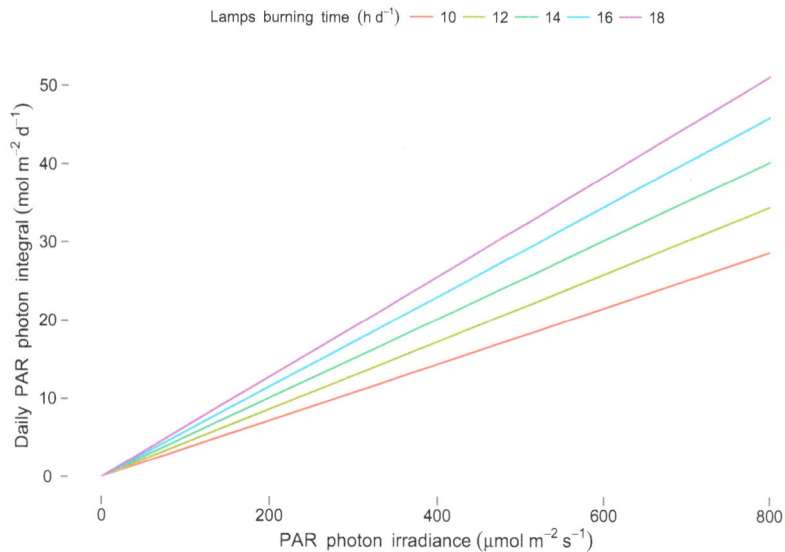

Figure 4.6: Daily PAR photon integral as a function of PAR photon irradiance supplied by lamps and the daily burning time of lamps. The values of interest are measured at the top of the canopy, and are applicable to supplemental light in greenhouses and total light in controlled environments.

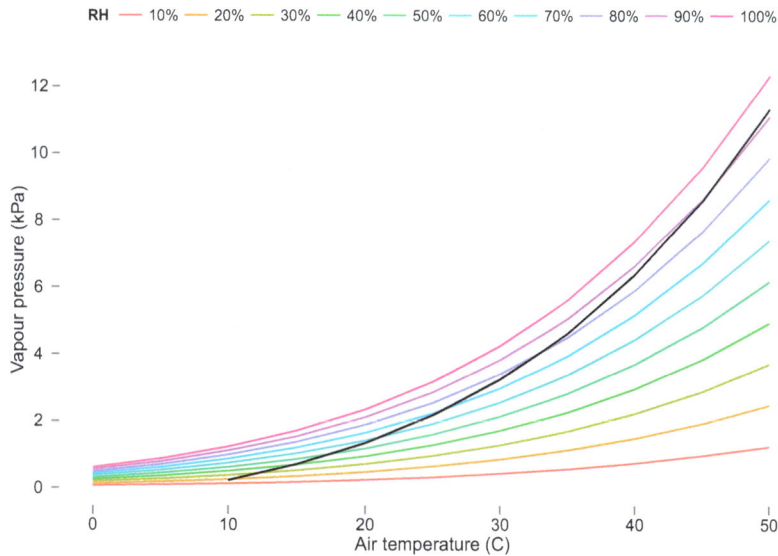

Figure 4.7: Water vapour pressure as a function of air temperature. A psychrometric chart showing the partial pressure of water vapour at different air temperatures and relative humidities. The driving force of transpiration from leaves is the absolute humidity gradient from the intracellular spaces out through the stomata to the surrounding air, called the vapour pressure difference (Δe), while the vapour pressure deficit (VPD) is the difference between actual and saturated water vapour pressure in the air. Unless leaf temperature equals air temperature Δe and VPD will differ. The black line is an example of a constant VPD = 1 kPa, assuming that the leaf temperature equals the air temperature this will also be the value of Δe. For example, 50% RH at 15°C gives the same VPD as 80% RH at 32.5°C and thus the same evaporative demand.

the absolute content of water vapour in air of the same RH is strongly temperature dependent (Figure 4.7), the evaporative demand can differ greatly when temperature fluctuates in the greenhouse, even if RH is stable.

In modern climate computers a measured quantity equivalent to VPD (i.e. based on absolute difference in humidity, expressed as water mass concentration, instead of difference in partial pressure) can be used for humidity control, instead of RH. It is called delta-χ ($\Delta\chi$, $g\,m^{-3}$, or sometimes simply delta-X) and is the water deficit from the saturated value, but expressed based on mass concentration instead of vapour pressure. Both VPD and $\Delta\chi$ use saturated water vapour content based on air temperature as a reference. To be correct we need information about the canopy temperature to use VPD or $\Delta\chi$ to estimate the true evaporative demand (Δe) as when the leaves are transpiring their temperature will be somewhat lower than the air temperature, due to evaporative cooling. The canopy temperature is not easily measured but in the context of climate control the approximation based of VPD or $\Delta\chi$ still provides a better approximation to the true evaporative demand created by the atmosphere in the greenhouse than RH.

Misting systems using ultrasonic humidifiers have been reported to also generate some hydrogen peroxide and may not be appropriate in research studies of plant growth (Arends et al., 1988). Although the evidence is not strong, one should be aware of this possible undesired effect.

However, increasing air humidity in greenhouses by forced aerosolization of water (creating small droplets by water sprays, jets, ultrasonic nebulizers etc.) may cause different problems. These processes cause charge separation of water droplets and thus forming high concentrations of nano-sized ions, which can damage electronic equipment due to electrostatic deposition. The effects of the charged droplets on plants are not known. This effect is not seen if water is being evaporated directly from a water surface to water vapour e.g. by trickling water in fans with cooling pads.

4.2.4 Elevated carbon dioxide concentration

Modern greenhouses have systems to supply elevated CO_2 concentration to promote crop growth. There are several possible sources of CO_2 that can be used: the three most common ones are liquid CO_2 from a tank, combustion of natural gas for CO_2 production and CO_2 produced when burning fossil fuels for heat/power production, where the latter two require cleaning of the fumes to remove NO_x and SO_x.

A major difference between elevated CO_2 supplied in full-scale greenhouses compared to open-top chambers or FACE is that the greenhouse CO_2 is not supplied continuously during daytime hours. Because of the vast volume of a greenhouse compartment it would be very expensive to supply elevated CO_2 when the vents are open for temperature control. Therefore in traditional greenhouse climate control elevated CO_2 is given only when the vents are closed or open by at most 5%. Ethylene contamination, which can significantly reduce plant growth, has been reported in some CO_2 supplies and it should be removed by $KMnO_4$ scrubbing if necessary (Morison and Gifford, 1984).

Furthermore the isotope composition may differ in liquid CO_2 compared to natural air, which need to be checked if investigations include measurements of the $^{13}C/^{12}C$ ratio e.g. in relation to water balance, and when comparing with results in the literature.

4.2.5 Growth substrates, irrigation and fertilization

Transferring experiments indoors inevitably means growing the plants in pots. It is important to chose a growth substrate that has both good water holding and draining properties to meet the demand for water and oxygen for the root system. In the horticultural greenhouse industry peat mixtures are the most commonly used substrate in pots but also coco-peat (granulated husks of coconuts) is used. To ensure good aeration perlite or vermiculite is sometimes mixed into the substrate. Since the pot volume is limited the fluctuations in water content will be greater than in soil, as will be the fluctuations in soil temperature. In natural soil the temperature will be both more stable and lower than what it is in pots.

The pots are often of dark colour and they absorb heat radiation when exposed to sunlight. In commercial production the pot density is kept high enough to create shade between the pots and only the outer row of pots are potentially exposed to heat radiation from the sun. In experiments you often leave some space between the plants, not to have excessive shade between them. During sunny days without shade the pot temperature can easily increase to levels that are damaging to the root system and when knocking the soil/root out of the pot of border plants that has been exposed to direct sunlight it is not uncommon to see root death on the exposed side of the pot. The temperature of the nutrient solution/water also affects the root temperature. If the water is cold a growth retarding effect can be seen on the plants. More reflections on pot experiments are found in Passioura (2006) and a meta-analysis of the effect of pot size on plant growth in Poorter, Bühler et al. (2012).

Several types of watering systems exist but the most common ones are drip irrigation and ebb/flow systems.

Drip irrigation is based on thin tubes supplying each pot with nutrient solution from a tank. In an ebb/flow system the nutrient solution it pumped up on the bench to 1-2 cm water depth and kept there for some minutes, before being drained off, back into the tank. This system recirculates the nutrient solution and with commercial nutrient/watering computers the composition is controlled by pH and electric conductivity (EC) of the solution, topping up with stock solution or an acid to keep the nutrient concentration and pH at desired level.

Irrigation of the plants can be done by timer or on demand. In greenhouses plants will have different need for water dependent on the weather. If the irrigation is done by timer without adjustment for sunny or overcast weather there will be a risk that the plants are overwatered during dull days or water stressed on sunny days. Both stresses potentially affect the growth of the plants.

Watering by demand can be done by experience or by weighing the plants and watering after a pre-determined water loss. Load cells can be mounted under ca. 0.5 m^2 watering trays or under a section of a greenhouse bench, triggering a pump in the tank with nutrient solution via a datalogger. Since plants that grow to maturity follow a sigmoid growth curve the watering needs to be adjusted accordingly during the experiment, to follow the demand.

4.2.6 Data logging of microclimatic variables

When working in greenhouses it is not enough to report the set points used for the most important climate parameters. The climate in a greenhouse needs to be measured and the observed values reported. The optimum arrangement is to have separate sensors recording the climate to those that are controlling the climate, but if this is not possible the ones connected to the climate computer can be used, if they are calibrated. Light irradiance, air temperature and humidity are the most important parameters to record, along with CO_2 concentration if it is applied. Since the natural irradiance can fluctuate rapidly, it is important to have a high logging frequency, e.g. every minute. Furthermore, it is important that the temperature sensor is shielded from direct heat radiation from the sun and the lamps, while still being exposed to free air, e.g. in a ventilated box. For ideas on how to construct a radiation shield for measurement of air temperature refer to the design of weather stations (e.g. `http://www.kippzonen.com`, `http://www.skyeinstruments.com` and may others). There are several small dataloggers on the market, with built-in air temperature and humidity sensors in a plastic housing. They also have to be shielded from heat radiation when used in a greenhouse or climate chamber since the plastic housing gradually heats up and transfers

the body temperature to the sensor—in climate chambers with metal halide lamps errors larger than 10°C have been observed when no shielding was used. Recommendations for sensors and frequency of data logging are compiled in Table 4.1.

4.3 Open top chambers and FACE

Open top chambers (OTC) and, less frequently, closed top chambers (CTC) are used to study the effects of elevated CO_2 concentration and temperature. Most of these chambers are built using plastic films, plates, or glass that absorb UV-B radiation and sometimes also some UV-A radiation. In some cases, attenuation of UV radiation has been minimised by use of UV-transparent materials in OTCs and CTCs e.g. Visser et al., 1997. In contrast to OTCs and CTCs, free air CO_2 enhancement (FACE) systems do not significantly affect UV or visible radiation. In some cases, UV-B lamps have been used in OTCs and the UV-B attenuation measured. In controls with unenergised lamps, daily integrals of UV-B and UV-A radiation, and PAR were reduced by 24% (Booker et al., 1992).

4.4 Controlled environments

In controlled environments the spectral distribution of the light emitted by the lamps used is especially important since there is no contribution from natural light. A skewed spectrum can cause strange growth patterns in plants and many lamp types produce distinct emission bands, interrupted by bands of very low emission (Figure 4.8). Fluorescent tubes and HPS lamps, in particular, emit very little far-red and produce plants with very stunted growth, compared to lamps that create a continuous spectrum resembling sunlight (Hogewoning et al., 2010). However, lamps emitting a continuous spectrum can also create unexpected differences in plants compared to sunlight. An unpublished example is that of white cabbage grown in pots outdoors then moved into a climate chamber with metal halide bulbs (see Figure 4.8), which developed a distinct red coloration after a couple of hours at constant light.

UV-B lamps emit some blue light providing a weak visual indicator that the lamp is turned on, but the contribution to the light environment will be minor in comparison to the overall PPFD from the main lamps used for PAR. They also emit very small amounts of orange-red radiation.

Most lamps presently used also emit a substantial amount of heat radiation. Some of this heat should be removed using a heat filter. The most efficient design to achieve this is to have the lamps in a separate, ventilated compartment in the ceiling, divided from the growth

Table 4.1: Guidelines for describing growth conditions in controlled environments and greenhouses, based on the recommendations of The International Committee for Controlled Environment Guidelines (http://www.controlledenvironments.org/) on which parameters should be recorded for proper description of conditions in climate chambers. We have modified it slightly and added recommendations for measurements in greenhouses (for the climate parameters also valid for them). In greenhouses and in the field a data logger with temperature, PAR and humidity sensors is required. n.a.: not applicable.

Parameter		Units	Where to measure	When to measure — Climate chamber	When to measure — Greenhouse	What to report
Radiation	PPFD	$\mu mol\,m^{-2}\,s^{-1}$	Top of the canopy in the centre of growing area (in greenhouses > 1 sensor is preferred)	Start, every 2 weeks and end of experiment	n.a.	Mean (daytime) and s.d. Radiation source (type, model, manufacturer)
	PPFD	$\mu mol\,m^{-2}\,s^{-1}$	As above	n.a.	Logging every minute	Mean max. daily PAR and s.d. Mean integrated PAR and s.d. Radiation source (type, model, manufacturer)
	Photoperiod	h	n.a.	n.a.	n.a.	Duration of light and dark period (in greenhouse for start and end of experiment).
Temperature	Air	°C	Top of canopy in the centre of growing area	Daily during each light and dark period, at least 1 h after light/dark change	Logging every 1–5 minutes	Mean and s.d. for day and night temperature, specification if it does not follow the photoperiod.
Air humidity	Water vapour pressure deficit (VPD)	kPa	Top of canopy in the centre of growing area	As for air temperature	Logging every 1–5 minutes	Average and standard deviation
	Δx	$g\,m^{-3}$	As above	As above	As above	As above
	Relative humidity	%	As above	As above	As above	As above (only if VPD or Δx not available)
Carbon dioxide	If elevated and part of the treatment	$\mu mol\,mol^{-1}$	Top of canopy	At least hourly	Logging every 1–5 minutes (via climate computer)	Mean and s.d.
Air velocity	If available and part of the treatment	$m\,s^{-1}$	At one or more representative canopy locations	At least once during the experiment		Mean and s.d.
Substrate	n.a.	n.a.	n.a.	At start of experiment	At start of experiment	Type and volume per container, components of soil(-less) substrate, container dimensions
Watering	n.a.	l (litre)	n.a.	Daily	Daily	Frequency, amount and type of water added
Nutrition or liquid culture	Acidity	pH	In bulk solution	Before and after pH correction	Before and after pH correction	Mean and s.d.
	Electric conductivity (EC)	$S\,m^{-1}$	As above	Before and after EC correction	Before and after EC correction	Average and standard deviation
	Ion composition	$mmol\,l^{-1}$	As above	Daily or when replenished	Daily or when replenished	Ionic concentration in initial and added solution. Aeration if any. Volume of initial solution.
	Solid media	$mol\,kg^{-1}$	n.a.	When added or replenished	When added or replenished	Nutrients and their form added to soil media.

chamber by a transparent glass/plexi-glass sheet. Even so, it is important that the temperature sensor measuring air temperature in a climate chamber is shielded from the heat radiation from the lamps. Water is a good absorber of infra-red radiation, and sometimes a transparent tray with circulating water[2] is used as a barrier between lamps and the plant compartment. Most lamp types (e.g. metal halide) used in climate chambers can only be turned on/off, not dimmed. To create different photon irradiances the lamps has to be turned on in different constellations. Even though daylight shows strong and fast variations in irradiance (Figure 4.10 on page 132) the overall light pattern is bell shaped. To mimic natural light conditions a gradual or stepwise change during the morning/evening hours is preferred. If it is possible to get e.g. $600 \, \mu mol \, m^{-2} \, s^{-1}$ from the lamps, a 14 h-long day will give an integrated PPFD of $30 \, mol \, m^{-2} \, d^{-1}$, which corresponds to a sunny day in May in Denmark (Figures 4.1 and 4.6). With the present development of the LED technology the first climate chambers with LED lighting have been introduced to the market, and one can expect that LEDs will be used extensivelly for controlled environments in the future. LEDs can be dimmed without any drastic changes in their spectral profile and the lack of heat radiation in the direction of the plants removes one complicating factor for lighting in climate chambers. Figure 4.8 shows the photon emission spectra of some lamps used as PAR sources in controlled environments and greenhouses.

In climate chambers it is often possible to create a fast change between the day and night temperatures, though this is unnatural to plants. Under natural conditions a 5–10°C decrease in air temperature normally takes hours, allowing e.g. the membrane fluidity to follow. If the temperature drop between day and night is big and fast some plants may experience transient stress linked to membrane fluidity, which would not be the case if the change were slower. This is something to bear in mind, when designing the chamber climate. It can also increase the risk of fungal disease if the air humidity is high and the plants become damp, in the same way as in greenhouses.

If experiments are done at low or high temperatures there may be restrictions on how well humidity can be controlled. At low temperature there may be limitations on the capacity for dehumidification and at high temperatures it may not be possible to sufficiently moisten the air.

Numerous plant species lose the ability to effectively regulate their stomata when grown at constant high humidity (ca. 70% RH at 20–25°C). To our knowledge, this has been observed in *Helianthus annuus*, *Lycopersicon esculentum*, *Cucumis sativus* and *Gerbera × hybrida*. These plants lose their ability to regulate stomatal conductance when subsequently exposed to low air humidity e.g. if moved from the greenhouse to a laboratory for measurements of photosynthesis. In the worse case scenarios, the plants transpire to such an extent that small local spots of necrosis develop on the patch where gas exchange has been measured on the leaf. However, since results like these are considered as failed experiments such observations have not (to our knowledge) been scientifically published. Nevertheless, it is important to be aware of this potential problem when working with mesophytic leaves, i.e. "average" leaves that are adapted neither to dry nor to wet conditions, like the examples above.

Data logging of the climate is needed in controlled environments to catch irregularities and failings in the climate control. This is especially important to test for differences among chambers when the climate set points changes. The Committee on Controlled Environment Technology and Use under the USDA has issued recommendations for data logging of growth chamber climate, which are compiled, with some changes, in Tables 4.1 and 4.2.

Recommendations on parameters that should be logged have been published by the International Committee for Controlled Environment Guidelines - see `http://www.controlledenvironments.org/`, `http://www.ceug.ac.uk/ICCEG.htm` and `http://www.controlledenvironments.org/guidelines.htm`.

4.5 Material issues in greenhouses and controlled environments

A potentially problematic issue concerning materials in greenhouses and climate chambers relates to the use of different types of soft plastic materials. Plastics like cellulose acetate, cellulose nitrate or PVC are stiff by nature. It is only when they are mixed with a softener i.e. a "plasticiser" that they become soft and flexible. The plasticisers may be different alkyl esters of phthalic acid. Some of these phthalates are phytotoxic, particularly butyl phthalate (DBP) and di-isobutyl phthalate (DIBP) (Hannay and Millar, 1986; Millar and Hannay, 1986). The phytotoxicity is evident as poor growth, chlorosis, necrosis and plant death. At regular intervals since the 1950's there have been examples in the greenhouse industry and plant research facilities where new plastics have been introduced for different purposes, which have caused serious growth problems (Hardwick and Cole, 1987). Many species are damaged by phthalates but species of Brassicaceae seem to be particularly sensitive. Past examples of products responsible for phytotoxic problems include plastic tun-

---

[2]Circulation or cooling of the water is needed to remove the heat from the system, otherwise the water could even boil.

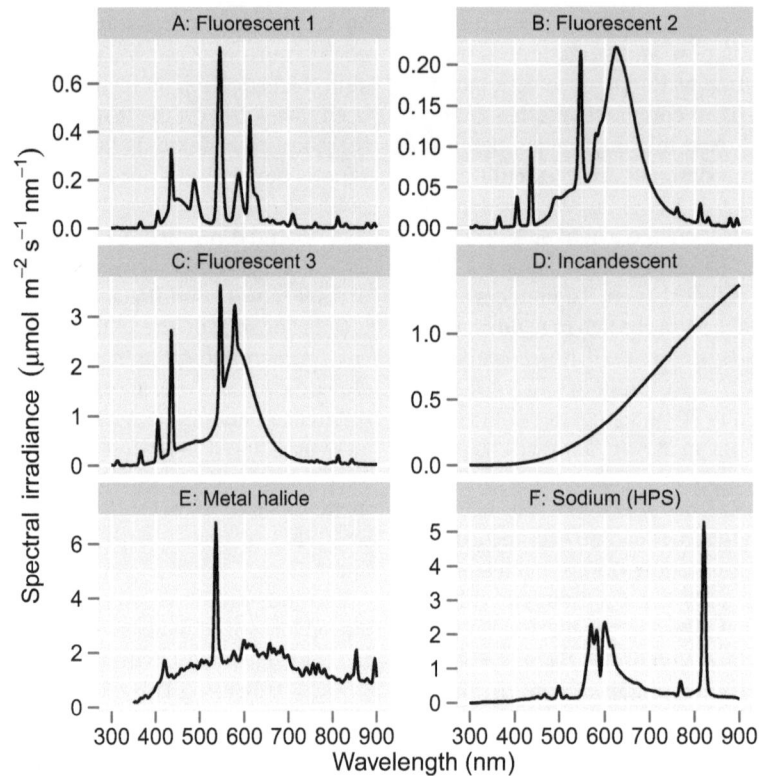

Figure 4.8: Photon emission spectra of visible light emitting lamps commonly used as PAR sources in controlled environments. A: T8 fluor. tube 'cool daylight' 36 W Philips TL'D' 36W/865; B: T8 fluor. tube 'incandescent' 36 W Philips TL'D' 36W/92; C: T12 fluor. VHO 'cool white' 215 W Sylvania F96T12/CW/VHO; D: 60 W incandescent lamp; E: Osram HQI 400 W/D metal halide discharge lamps; F: Osram NAV-T400 high pressure sodium lamp. Lamps have been measured individually and in banks, and the distance to the spectroradiometer differs, meaning that absolute spectral photon irradiances are not comparable between panels. Measured with a LI-1800 spectroradiometer (LI-COR, Lincoln, NE, USA).

Table 4.2: Guidelines for describing controlled environments and greenhouses. n.a.: not applicable.

Parameter	Units	What to report
Manufacturer	n.a.	Name and address (n.a. for greenhouse)
Model	n.a.	Model descriptor if available. Greenhouse type and roof height for greenhouses
Size	m^2	Floor area
Barrier beneath lamps	Yes/No	Indicate if present and composition. Optical properties if available (not greenhouses)
Cladding material	n.a.	Indicate material and optical properties if known (only for greenhouses)
Air flow	n.a.	Indicate whether up, down or horizontal (not greenhouses)

nels, water hoses, drip irrigation tubes, plastic plant pots, glazing strips (that seal between the glass and aluminium frames in the greenhouse construction), plastic insulation of electrical wires and plastic boots—the latter creating growth problems in a greenhouse outside a staff changing room. Many of these problems have been created when the recipes or composition of products already in use in a greenhouse have been changed by the manufacturer. Growth chambers require some fresh air inlet for the climate control. There is an example of poor growth and bleaching of plants in a climate chamber when the air inlet was close to the ventilation outlet of a chemistry laboratory. When the air inlet was moved 200 m away and supplied with a compressor, the problem disappeared.

4.6 Gas-exchange cuvettes and chambers

Measurement of photosynthesis through gas-exchange with UV-B present requires modifications of most standard equipment. Most gas-exchange cuvettes have windows that attenuate UV radiation. This means that most leaf photosynthesis and stomatal conductance measurements are done in the absence of UV radiation, even when done in sunlight. Stomata have been shown to respond to UV radiation (e.g. Eisinger et al., 2000), and at high UV irradiances the rate of photosynthesis can also be affected (see Rozema et al., 1997, for a consice review).

For example, the GFS-3000 system from Walz, can be ordered with cuvette windows made of quartz instead of normal glass. Some cuvettes for the LI-COR LI-6400 use a plastic film as a window, and it is possible to use an UV-transparent material. The cuvette for CIRAS-2 from PP Systems is delivered with Calflex heat filter glass as standard, but can easily be fitted with quartz glass for UV-B transmission on request.

Frequently used light sources are tungsten-halogen lamps, which emit very little UV radiation, or light sources based on red LEDs in combination with blue or white LEDs, which do not emit any UV-B radiation.

When designing experiments, especially when leaves or shoots are enclosed for more than a couple of minutes in a gas-exchange cuvette, one should be aware that the leaves or shoots may be in an unnatural light environment, even when measurements are done in full sunshine. To avoid surprises, always measure irradiance and especially spectral irradiance through the cuvette window, rather than outside the cuvette, if you want to describe the radiation environment during measurements.

Plastic and rubber used for tubing have different properties of permeability to CO_2 and water vapour, and absorption of water, which can lead to flawed gas-exchange data. These properties are listed in Long and Hällgren (1987).

4.7 Plants in the field

One important disturbance in experiments on natural vegetation or field crops is that produced by the researchers themselves. Some ecosystems like peat bogs get more easily damaged than others, but in most cases special precautions are needed whenever repeated access to plots is needed. The most common approach is to use catwalks made from wooden planks raised some centimetres above the ground. Sometimes when access is needed to the centre of the plots ladders are put temporarily, laying horizontally bridging two catwalks. Figure 4.9 shows some examples from Abisko and Figure 2.24 on page 56 shows an example from Ushuaia.

Not only the spectrum, but also the temporal variation of irradiance differs between controlled environments and natural conditions. The temporal light regime depends on the position of the sun and on clouds (Figure 4.10), and also on the vegetation itself (Figure 4.11). For experiments done in the field summaries of data from an in situ or nearby weather station should be used to describe the growing conditions. To assess the light environment at different locations in a forest understorey, hemispherical photographs taken with a fish-eye lens, can be used. There is software available—e.g. Hemiview from Delta-T—that allows the prediction of the light environment based on the position of the sun in the sky at different times of the day and seasons of the year.

4.8 Cultivation of aquatic plants

When designing experimental set ups for the cultivation of aquatic plants under full solar radiation (PAR + UV-A + UV-B) or under a simulated natural radiation field (see section 2.4 on page 62), the spectral transmission of the vessels or photobioreactors used must be accounted for as well as the absorbtion of UV radiation by the water, growth medium and the aquatic plants themselves (self shading). Most studies evaluating photosynthetic activity in algae have been conducted in UV-opaque incubators and this has consequently led to overestimation of their photosynthetic capacity, since UV radiation causes photoinhibition and photodamage in aquatic plants (Bischof et al., 2006; Häder and Figueroa, 1997; Villafañe et al., 2003).

The most frequently used photobioreactors have a cylindrical shape and they can be transparent or opaque to UV radiation (Figure 4.12, see section 2.4 on page 62). Some photobioreactors are built from rigid plastic such as polycarbonate (PC), polymethylmethacrylate (PMM) or polyvinyl chloride (PVC), and others from flexible plastic such as cellulose acetate or polyethylene.

The aims of photobioreactor design are: (a) high volu-

Figure 4.9: View of two UV-B experiments in Abisko, Sweden, showing the use of catwalks to protect the natural vegetation under study. Left: use of small UV-B lamps on small plants; Dr. Carlos L. Ballaré. Right: Use of frames with 40W UV-B lamps; Prof. Lars Olof Björn. Photographs: Pedro J. Aphalo.

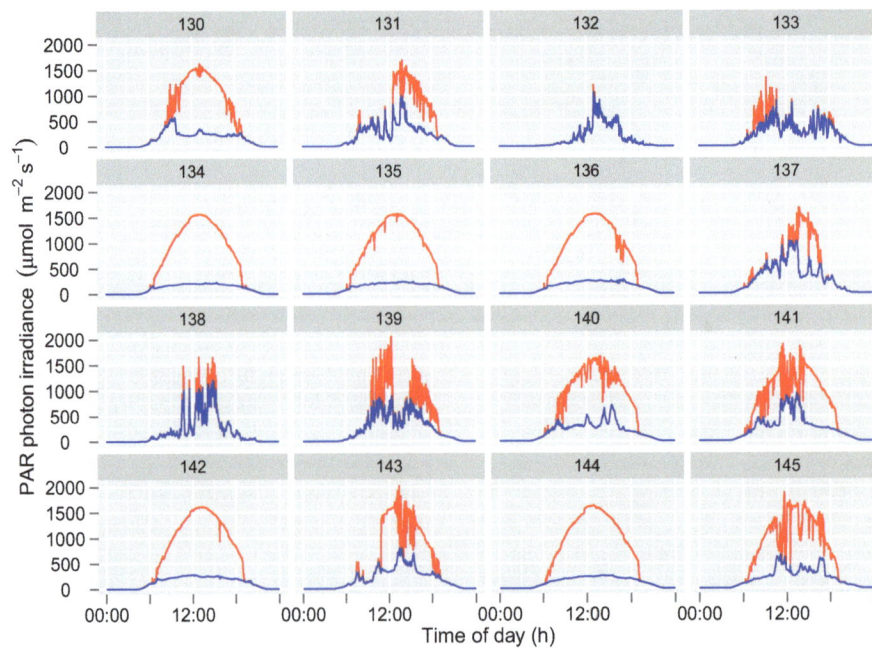

Figure 4.10: The effect of clouds on light regime. Total (red) and diffuse (blue) PAR photon irradiance throughout 16 Spring days, starting from day 130 or 9 May 2012, in Viikki, Helsinki, Finland. Measured with a BF5 instrument (Delta-T, Cambridge, U.K.) and a CR10X datalogger (Campbell Scientific, Ltd, Logan, Utah, U.S.A.). Plotted data are one minute averages. Panel headings indicate the day of year. Unpublished data: Pedro J. Aphalo, T. Matthew Robson, Oriane Loiseau and Saara Hartikainen.

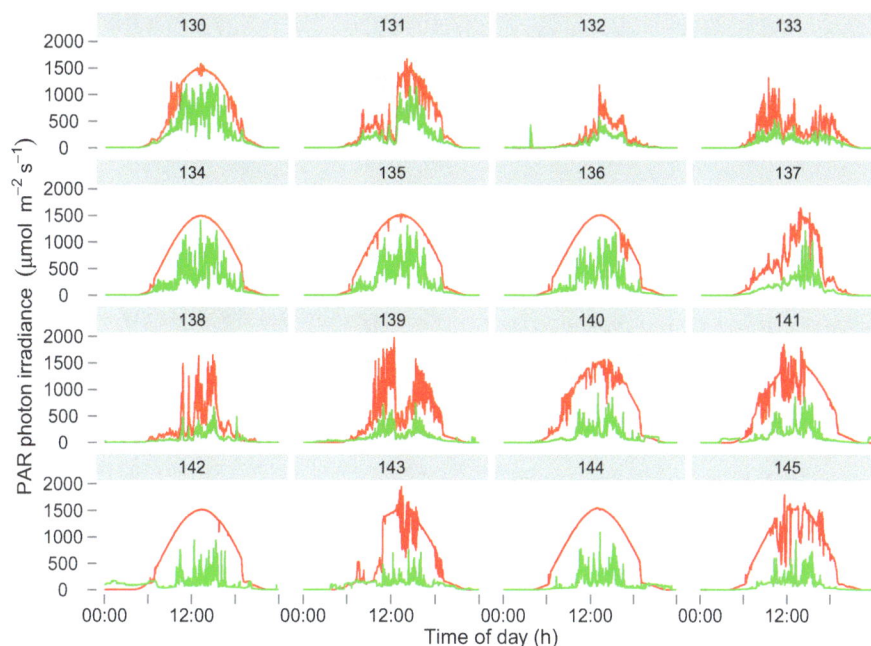

Figure 4.11: The effect of vegetation on the light regime. Total PAR irradiance in the open (red) and in the understorey of a grove of silver birch saplings (green). Panel headings indicate the day of the year, starting from day 130 or 9 May 2012. Bud break took place on 24 April 2012, day 115 of the year. Viikki, Helsinki, Finland. Measured with two LI-190 quantum sensors (LI-COR, Lincoln, NE, U.S.A:) and a CR10X datalogger (Campbell Scientific, Ltd, Logan, Utah, U.S.A.). Plotted data are one minute averages. Unpublished data: Pedro J. Aphalo, T. Matthew Robson, Oriane Loiseau and Saara Hartikainen.

Figure 4.12: Different culture systems for research on the effect of solar and artificial radiation on macroalgae. (A) UV-transparent Polymethylmethacrylate (PMM) cylinders; (B) Non UV-transparent Polycarbonate (PC) cylinders; (C) Algae grown in non-UV-transparent polyethylene (PE) bags ; (D) Glass walled aquaria with macroalgae, are not transparent to UV radiation.

metric productivity (g l^{-1} d^{-1}), (b) high productivity per unit area (g m^{-2} d^{-1}) (c) high cell concentration (g l^{-1}) (d) high efficiency in the conversion of light (g mol$^{-1}_{photons}$) (e) high biomass quality (f) sustainable and reliable cultivation, and (g) low construction and operating costs.

The material for the photostage of the photobioreactors must have: (a) high transparency, (b) high mechanical strength, (c) high durability (resistance to weathering), (d) chemical stability (e) ease of cleaning, and (f) low cost.

Seaweeds can be also cultivated in large tanks (500–1000 l) made of materials such as polyethylene, glass-fibre-reinforced polyester or polypropylene—which all attenuate UV radiation (Figure 4.13). Thus when UV and visible radiation are supplied by artificial lamps, they must be located above the open top of the tanks. Likewise when tanks receive natural solar radiation, direct radiation is available within the tank only at high solar elevation angles, whereas scattered or 'diffuse' radiation predominates during the rest of the day. Large containers are considerably more expensive than other systems and require an ample water supply and frequent maintenance. Facilities originally designed for other purposes, e.g. biofiltration of fishpond effluents, are often used for the incubation of aquatic plants (Figueroa et al., 2006; Neori et al., 1996). These tanks can be placed outdoors under full solar radiation or enclosed inside a greenhouse.

In tanks, incident radiation is generally attenuated very rapidly and attenuation depends largely on algal density, the shape and volume of the tanks and the amount and type of particulate and dissolved material in the water. If we compare the irradiance at different depths in the water with different regions of the light-response curve of photosynthesis, we can see that PAR irradiance can be high enough to cause photoinhibition in the surface layer (Figure 4.14). The smallest irradiance causing photoinhibition is denoted in the graph as Q_h while the irradiance incident at the surface of the water is Q_0. In the next layer below the surface, the incident irradiance is lower than Q_h but higher than saturating irradiance for photosynthesis (Q_s). About 80–90% of photons reach this layer of maximal production. The irradiance in the layer below, which is deeper than the layers above, is lower than saturating irradiance but higher than light-compensation-point irradiance (Q_c)[3]. Finally the irradiance in the deepest layer at the bottom of the tank is lower still Q_c, this is considered to be a 'dark' layer (in general 70% of the volume of the tank). Thus, it is crucial that the algae are vigorously circulated through the tanks using air to maintain a light:dark regime adequate for growth, i.e. the algae move from the bottom of the tanks, where the tubes injecting the air are located, to the upper layers, where photon

irradiance is higher.

To cultivate both fixed or free floating aquatic plants, the plants must receive air (with or without CO_2 enrichment) in addition of light and nutrients. Air not only serves as a carbon source but also produces the turbulent environment needed to provide nutrients to the algae, which under a laminar regime would be less available to some of the algae leading to their starvation. Thus, the suitable movement of macrophytes within the tank results, not only in their adequate exposure to PAR and UV radiation, but also the provision of nutrition and gas exchange. Thus, the appropriate balance between the initial inoculum and the water volume should be accurately calculated. Due to circular movement of water within a tank, macrophytes are often exposed to variable light conditions or, when densities are high, to self-shading. The cultivation of subtidal seaweeds in outdoor tanks can cause bleaching or chronic photoinhibition because they are not normally exposed to high irradiances of PAR. These effects of high irradiance can mask the effects of UV radiation. This can be avoided by the use of neutral shading screens that minimise the undesirable effects of excessive light, as exemplified in a tank cultivation of three seaweeds from the coast of Baja California (Cabello-Pasini et al., 2000). By neutral screens, we mean grey or black shade-cloth that has little or no effect on the radiation spectrum, attenuating all wavelengths by the same relative amount. When using more than one layer of shade-cloth, remember that the effect is multiplicative rather than additive. For example two layers of a cloth with a 50% transmittance each will yield a transmittance of approximately 25%.

As an example, in cultures of *Falkenbergia rufulanosa* in semitransparent polyethylene tanks, the high algal biomass density caused a drastic reduction in the maximal average irradiance at 10 cm depth compared to that at the surface (Figueroa et al., 2008). At 10 cm depth, PAR was reduced by about 87.5% at an algal density of 4 g l^{-1}, 89.5% at 6 g l^{-1} and 95% at 8 g l^{-1}. The attenuation of UV-A radiation was 94% at 4 g l^{-1}, 96% at 6 g l^{-1} and 99% at 8 g l^{-1}. UV-B radiation at noon was fully attenuated by an algal density of 4 g l^{-1} at 10 cm depth, while with this same algal density there was "complete darkness" (less than 0.1% of incident PAR irradiance) at 25 cm depth.

The shape of the tank and the ratio of surface to volume (S:V) also have an effect on the hydrodynamics and the movement of the seaweeds (Figure 4.15). Semicircular tanks of fibre glass (750 l) have better hydrodynamics, not only because they allow more water movement and a better distribution of the air bubbles in the tank, but also because of their higher S:V ratio (2.4 m^{-1}) than

---

[3]At the light compensation point gross photosynthesis and respiration rates cancel each other, and the rate of net carbon exchange is equal to zero.

Figure 4.13: Large volume tanks used for incubation of macroalgae. Different tank systems at the Center for Marine Biotechnology at Las Palmas G.C. University (Gran Canaria, Spain) both indoor and outdoor (A-C); and at the National Center for Mariculture (Israel Oceanographic and Limnological Research Ltd, Eilat, Israel), outdoor (D) (Photographs by Félix L. Figueroa)

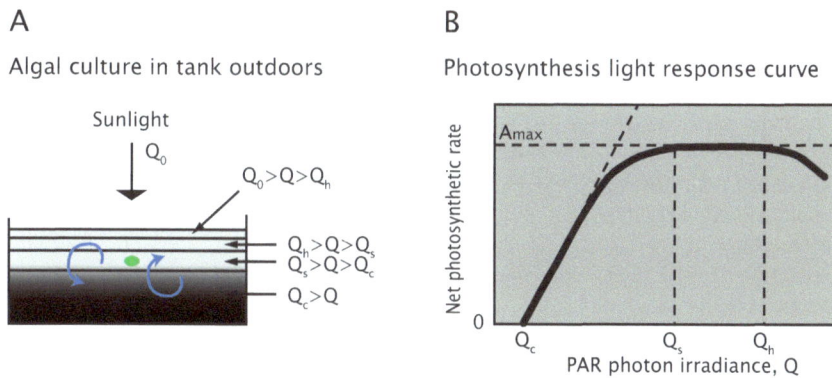

Figure 4.14: (A) The typical pattern of change in irradiance with depth in a tank; layers correspond to the irradiance response curve of photosynthesis (B). Q_0 is the incident photon irradiance at the surface, Q_c the light compensation point (the photon irradiance at which gross photosynthesis and respiration cancel out, Q_s the photon irradiance for saturated photosynthetic rate (stationary phase of photosynthesis versus irradiance) and Q_h is the smallest irradiance producing photoinhibition. The photon irradiance Q is below Q_c in approximately 70% of the volume of the tank, but because of water movement the algae move between the layers. (Redrawn from diagrams by Liliana Rodolfi, used with permission.)

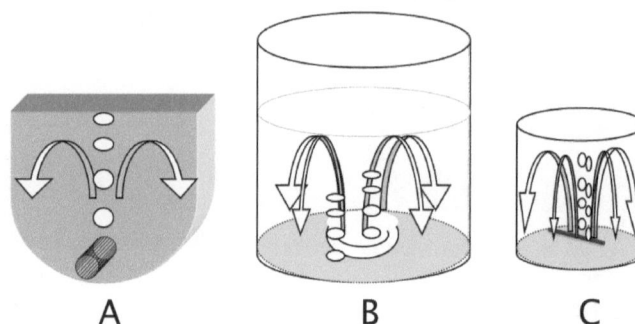

Figure 4.15: Examples of tanks used for experiments with macroalgae, (a) semicircular tank of 750 l surface/volume ratio (S/V)= 2.4 m^{-1} , (b) circular tank of 1500 l of capacity , S/V ratio=1 m^{-1} and (c) circular tank of 90 l , S/V =2.2 m^{-1}).

that of cylindric 1500 l tanks of semitransparent polyethylene (S:V = 1 m^{-1}) or 90 l (S:V = 2.2 m^{-1}).

Another aspect of good experimental design is choosing the location of plants within the culture system, i.e. the merits of fixed- versus free-floating aquatic plants. Both fixed and free floating aquatic plants absorb radiation to differing extents depending on their pigment composition and the thickness of their tissue or thallus. Aquatic angiosperms can be anchored to both natural materials, such as stones, and artificial materials (rows, epoxy, etc.). Seaweeds are usually cultured in free floating systems where they are moved by the injection of air, although in nature they are fixed to different substrates. Not all species can be cultivated floating freely and in some cases this can alter their growth forms, transforming branched forms into a spherical forms. Morphological changes affect both bio-optical properties (i.e. including UV absorption) and nutrient assimilation which is higher in branched forms than spherical forms due to their higher surface:volume ratio (of both cells and the whole plant). Thus, in photobiological experiments, it is crucial to optimise both algal density and the light field—by using lamps located to one side (lateral position) or above the aquarium—so as to reduce or avoid self shading.

Mesocosms have been developed to recreate natural conditions as closely as possible on a small scale. They are generally used to study plankton communities, however, their use in aquatic macrophyte ecology is increasing. For example, floating mesocosms and mesocosms connected to land—e.g. attached to a platform or pier—have been developed (Figure 4.16).

Several mesocosm studies have used ambient radiation supplemented with UV-B radiation from fluorescent lamps to simulate an increase in UV-B radiation due to ozone depletion. Two methodologies can be applied: (a) supplementing a fixed irradiance value ("square wave systems"), or (b) supplementing a fixed percentage of ambient irradiance ("modulated systems") (Belzile et al.,

2006; Belzile et al., 1998). Modulated systems, which were originally developed for terrestrial plants (Caldwell et al., 1983), are better than square-wave systems for simulating predicted increases in UV radiation (S. Díaz et al., 2006, 2003). See section 2.2.6 on page 48.

When using mesocosms, experimental designs must be as realistic as possible in simulating forcing factors as well as their variability. For example, when studying the effect of increased UV, experiments should be designed to reproduce diurnal variations in irradiance and episodic events such as passing clouds. Recently, experiments using mesocosms have been applied to evaluate the interactive effects on macrophytes of environmental shifts driven by climate changes (temperature, PAR, UV radiation and nutrient availability). To mimic climate change conditions, Liboriussen et al. (2005) designed a flow-through mesocosm in a shallow lake which can also be applied to other aquatic ecosystems. Recently, a new automatically-operated system providing accurate simulations of the increases in UV-B radiation and temperature according to climate change scenarios has been developed (Nouguier et al., 2007). All these systems reproduce the sort of high degree of environmental variability inherent to aquatic systems.

4.9 Recommendations

In this section we list recommendations related to the cultivation of plants used in experiments studying the effects of UV radiation. See sections 2.6.1 and 2.6.2 for recommendations on manipulation of UV radiation, section 3.15 for recommendations on how to quantify UV radiation, and section 5.12 for recommendations about statistical design of experiments.

1. Include in the methods description in your publications enough information about the growing

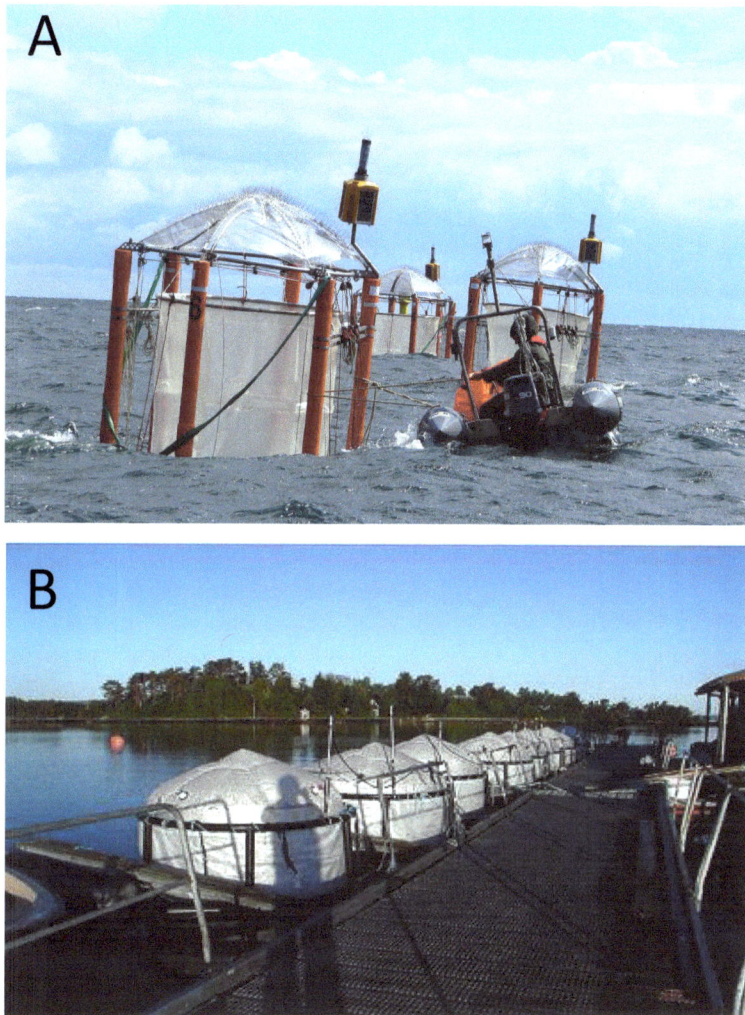

Figure 4.16: (A-B) Free-floating mesocosms of the Helmholtz Centre for Ocean Research Kiel (GEOMAR), Germany and (C) meso-cosms attached to a platform in Espeland Marine Biological Station, Bergen, Norway. Photos: (A) Ulf Riebesell, GEOMAR; (B) Jens Christian Nejstgaard, University of Bergen.

conditions so that the experiment can be duplicated.

2. When using growth chambers, growth rooms, or greenhouses with environmental control systems report in addition to model and manufacturer and its address, all the settings used:

    a) Temperature i. Day temperature (°C) ii. Night temperature (°C) iii. The temperature ramp (°C h^{-1}) used when increasing/decreasing the temperature. (This function is important in decreasing the risk of fungal disease due to moist plants, i.e. a too steep temperature drop will increase the risk of condensation on the plants.) iv. Temperature when ventilation starts (°C) v. Temperature when ventilation stops (°C) vi. Light-dependent temperature increase, if dynamic climate is applied

    b) Air humidity i. Upper limit of humidity where humidity control starts

    c) Lamps. i. Indicate type, and manufacturer, power rating. For rooms and chambers indicate whether they are separated from the plants by a barrier made of glass, acrylic, polycarbonate, or some other material. ii. For greenhouses indicate the ambient light set point when the lamps are turned on, and the ambient light set point when the lamps are turned off. iii. The global radiation (W m^{-2})

sensor is often used and recalculated to PAR values. Note that you can have a PAR (400–700 nm) radiation sensor in $W\,m^{-2}$, or $\mu mol\,m^{-2}\,s^{-1}$.

d) Shade/insulation screens. i. Specify screen type and ii. Irradiance ($W\,m^{-2}$) set point when the screens are closed. iii. Irradiance ($W\,m^{-2}$) set point when the screens are opened.

e) CO_2 supply i. CO_2 level when the vents are closed. ii. % ventilation with maintained CO_2 supply (how much open the windows are allowed before the CO_2 supply is stopped).

f) Pots, soil and fertilization. i. Indicate pot type and volume, or diameter and height. ii. Describe the soil in field experiments, and the composition when using artificial growth substrates. iii. Indicate type, and composition of fertilizers applied, and timing of applications.

g) Watering. i. Describe watering regime in controlled environments and greenhouses. ii. Indicate amount of rainfall in field experiments.

h) Plant material i. Indicate Latin name of species studied, and cultivar if relevant. ii. Indicate the source of seeds or vegetative multiplication material used.

3. Check the calibration of the climate sensors that control the greenhouse or controlled environment. For example, PAR sensors should be calibrated once a year, infrared gas analysers (IRGAs) used in CO_2 control once every week or two (but depends on the instrument, so follow the manufacturer recom-

mendations). Electronic thermometers and capacitive air humidity sensors (such as Vaisala HUMICAP) should be checked a few times per year, and recalibrated when needed.

4. Program the data logger for logging each minute, either a separate logger with sensor or the climate logging function of the climate computer using the same sensors as are controlling the climate. If you use a separate battery-powered datalogger, the data record will include the time during mains power interruptions or environmental control system failure.

4.10 Further reading

Poorter, Fiorani et al. (2012) have written a practical guide on how to grow plants for reproducible results, giving several examples of how variation in the climate influence the results, including suggestions for further reading. If biomass allocation to different plant organs is studied in an experiment, the Tansley review by Poorter, Niklas et al. (2012) should consulted as it presents a meta-analysis of how climate parameters other than UV-B affect plant morphology. This is especially important if results from different growing seasons are compared. For general introduction to the energy, water and carbon balance of plants in relation to the climate we refer to Jones (1992) and Nobel (2009), for climate physics in relation to plant canopies to Oke (1988) and for more thorough calculations of the greenhouse climate to Bakker et al. (1995); Beytes (2003); Stanghellini (1987); Stanhill and Enoch (1999).

5 Statistical design of UV experiments

Pedro J. Aphalo, T. Matthew Robson, Harri Högmander

5.1 Tests of hypotheses and model fitting

The guidelines given here for good experimental design have been formulated with particular consideration given to UV manipulation experiments. Nevertheless most of the principles behind these guidelines can be generally applied to the design of experiments.

Experiments can be designed to test hypotheses or to estimate the values of parameters in a model. In the first case, we should be careful not to let our knowledge of the data influence the tested hypothesis[1]. In the second case, model choice may to some extent depend on the data, but we should be aware that this *a posteriori* choice of a functional relationship, could make the P-values less strict for testing treatment effects.

Fitting models is especially useful when the parameters of the model have a biological interpretation. In this case, it is important to obtain estimates of the reliability of your estimates of the values obtained for the parameters. In other words, we need to present not only the fitted values, but also their confidence intervals, or at least standard deviations.

All estimates of P-values are based on a comparison of the variation between differently treated experimental units, and the variation among equally treated experimental units[2]. If we assign treatments objectively by randomization, we can expect that in the absence of a treatment effect, on average, there will not be more variation among differently treated units than between equally treated units. The quality of the estimates obtained for the magnitude of these sources of variation will depend on the number of replicates measured, on proper randomization, and on an adequate design being used to control known error sources. The remaining unaccounted variation among equally treated units is called *experimental error*.

When using tests of significance one should be aware that the P-value will depend on both the size of the treatment effect and on the standard deviation of the estimate (e.g. standard error of the mean) which strongly depends on the number of replicates. In an experiment with hundreds of replicates one is very likely to detect treatment effects that are statistically-significant, even if these effects are too small to be of any biological significance. On the contrary, in an experiment with few replicates, one risks biologically-significant effects remaining undetected. One should never forget that a large P-value indicates only that we have been unable to demonstrate an effect of a treatment, rather than demonstrating that such an effect did not exist. Strictly speaking it is impossible to demonstrate that there is no effect. However, what we can do is to estimate the size of the smallest effect that our experiment could have detected. This can be done by means of 'statistical-power analysis' (see Cohen, 1977; Quinn and Keough, 2002, Chapter 7).

John W. Tukey (1991) has written:

> Statisticians classically asked the wrong question—and were willing to answer with a lie, one that was often a downright lie. They asked "Are the effects of A and B different?" and they were willing to answer "no."
>
> All we know about the world teaches us that the effects of A and B are always different—in some decimal place—for any A and B. Thus asking "Are the effects different?" is foolish.
>
> What we should be answering first is "Can we tell the direction in which the effects of A differ from the effects of B?" In other words, can we be confident about the direction from A to B? Is it "up," "down" or "uncertain"?

---

[1] Ideally, if the data suggests a new hypothesis, this new hypothesis should be tested in a new experiment, not using the same data with which hypothesis was formulated. Alternatively, one can use a procedure to adjust P-values, as recommended for multiple comparisons.

[2] It follows that if we miss-identify the experimental units, the statistical tests are invalid.

The third answer to this first question is that we are "uncertain about the direction"—it is not, and never should be, that we "accept the null hypothesis."

5.2 Planning of experiments

It is very important to properly design and plan an experiment in advance of its execution. From a well planned experiment one gets more interpretable and reliable results, usually with less effort and expense, and even sometimes faster. Often, badly planned experiments can yield unreliable data or results that do not answer the intended objective of a study. There are too many examples in the scientific literature of badly designed experiments or invalid statistical analyzes leading to erroneous conclusions, and manuscripts submitted for publication are frequently rejected by journals on the basis of a poorly thought through design.

5.3 Definitions

Quoting Cox and Reid (2000):

Experimental units: are the patients, plots, animals, plants, raw material, etc. of the investigation. Formally they correspond to the smallest subdivision of the experimental material such that any two different experimental units might receive different treatments (e.g. filter frames, lamp frames, growth chambers).

Treatments: are clearly defined procedures one of which is to be applied to each experimental unit.

Response: the response measurement specifies the criterion in terms of which the comparison of treatments is to be effected. In many applications there will be several such measures.

See Box 5.1 for examples.

5.4 Experimental design

As discussed by Cox and Reid (2000), the most basic requirement for a good experiment is that the questions it addresses should be interesting and fruitful. Usually this means examining one or more well-formulated research hypothesis. In most cases the more specific the research question asked, the greater the chance of obtaining a meaningful result. Quoting J. W. Tukey "An approximate answer to the right problem is worth a good deal more than an exact answer to an approximate problem."

Once the research hypothesis to test has been chosen: 1. the experimental units must be defined and chosen, 2. the treatments must be clearly defined, 3. the variables to be measured in each unit must be specified, and 4. the size of the experiment, in particular the number of experimental units, has to be decided.

When we perform an experiment, we choose our experimental units and manipulate their environment by controlling the levels of a component factor within that environment (*treatments*). We then record the *response* of our experimental unit. *Factors* are groups of different manipulations of a single variable. Each of the distinct manipulations within a factor is called a level: e.g. in an experiment giving plants daily exposures to 0, 5, and 10 $kJ\,m^{-2}\,d^{-1}$ of $UV^{GEN(G)}$, the factor is $UV^{GEN(G)}$ and it has three levels (0, 5, and 10 $kJ\,m^{-2}\,d^{-1}$). A factor can also be qualitative, e.g. a factor where the levels are different chemicals.

Questions detailing the steps:

1. What is the purpose of the experiment? To which objective questions do we want to find answers?

2. What treatment is to be applied? At what levels? Do we include an untreated control? Is there any structure within the treatments?

3. What is the response to be observed and recorded? What is the nature of the observations?

4. Are there other variables which could affect the response?

5. How big a difference in response is practically important? How big a difference in response should be detectable?

6. How many experimental units are appropriate and practical to use?

7. How do we organize the experiment? What, where, when, how, who...?

8. How will the data obtained be analyzed?

In addition to performing these steps, keep notes in a logbook of the experimental plan, and everything done during the experiment. This will allow you to go back and check if the design was sound and how faithfully it was followed. By noting any changes that were made we may remember something important to assist in the interpretation of the results, or an improvement to make to future experimental designs. If necessary, such notes will also allow for any future repetition of the whole experiment. See examples in Box 5.2.

Box 5.1: Treatments, experimental units, and responses

- We have 100 plants, we apply pure water to 50 plants, and a fertilizer dissolved in water to the remaining 50 plants. After two weeks we measure the dry weight of all plants.

- The plants are in individual pots, so each plant might receive fertilizer or not: the plants are the experimental units.

- The manipulation procedures applied to the plants are: 1) apply water, or 2) apply water + fertilizer. These are the two treatments, which could be called 1) control, 2) fertilization.

- The response is the criterion we will use to compare the treatments: the dry weight of the plants two weeks after the treatments are applied.

Box 5.2: Design of an experiment with UV-absorbing filters: example

1. Purpose: study the effect of solar UV radiation on the accumulation of flavonoids in silver birch seedlings.

2. Response observed: concentration of flavonoids in the upper epidermis of the leaf. Nature of observations: epidermal absorbance measured at 365 nm wavelength with a Dualex FLAV instrument. Sequential measurements once a week.

3. Treatment: Solar UV attenuation using filters. Levels 10%, 50%, and 90% attenuation of $UV^{GEN(G)}$.

4. Other variables which could affect the response: age of seedlings, age and size of sampled leaf, soil type, temperature, rainfall, shading, time of application.

5. Number of experimental units: e.g. 5. Estimated based on item 9 below.

6. Number of seedlings (subsamples) measured in each experimental unit: e.g. 20.

7. How do we organize the experiment? What seed provenance, size of pots, sowing date, fertilization date if any, watering frequency, filter frame size and height, soil used, how frequently to sample and when, who does all these things.

8. Data analysis: compute the mean epidermal absorbance for each experimental unit and date, and test for differences between treatments in an ANOVA for repeated measurements, accounting for any uncontrolled environmental gradients (as blocks or covariates).

9. Difference in response that is practically important from a biological perspective: e.g. 0.3 absorbance units. Detectable difference: e.g. 0.2 absorbance units.

5.5 Requirements for a good experiment

Following D. R. Cox (1958, Chapter 1):

Precision. Random errors of estimation should be suitably small, and this should be achieved with as few experimental units as possible.

Absence of systematic error. Experimental units receiving different treatments should differ in no systematic way from one another.

Range of validity. The conclusions should have a wide range of validity.

Simplicity. The experiment should be simple in design and analysis.

The calculation of uncertainty. A proper statistical analysis of the results should be possible without making artificial assumptions.

In practice we almost never have too many replicates in UV experiments, as the responses under study tend to be small compared to the random variation. For UV experiments, more frequently the problem at hand is how to make the most efficient use of the limited number of true replicates that we can afford. Good experimental design helps with this and helps us to assess the feasibility of experiments given a limited amount of resources. Fulfilling the requirements listed above will ensure that conclusions derived from experiments are valid.

5.6 The principles of experimental design

We will consider three 'principles' on which the design of experiments should be based: replication, randomization, and grouping into blocks. Each of these principles ensures that one aspect of the design of an experiment is correct and increases the likelihood that it will yield data that can be statistically analyzed without need for unrealistic assumptions (Figure 5.1). In the sections below we discuss each of these principles in turn and give examples of their application.

5.6.1 Replication

When an experiment is done several times, either simultaneously or sequentially, each time or repetition is a *replicate*. Replication serves several very important roles. (a) It allows for estimation of the random variation accompanying the results. This random variation is called *experimental error*. (b) Replication increases the precision with which the treatment effect is estimated, because each repetition gives additional information about the experiment. This can be appreciated from the equation

relating the variance of the original population to that of the mean: $\sigma_{\overline{x}}^2 = \frac{\sigma^2}{n}$, where $\overline{x}$ is the mean, n the number of replicates, and where σ^2 is the variance (of individual measurements) and $\sigma_{\overline{x}}^2$ is the variance of the mean. The standard error of the mean is $\sigma_{\overline{x}} = \sqrt{\sigma_{\overline{x}}^2} = \frac{\sigma}{\sqrt{n}}$ frequently abbreviated to s.e. or S.E.M.

5.6.2 Randomization

Which treatment is applied to each experimental unit should be decided at random (if appropriate after forming blocks). The idea is to be objective. One should avoid any subjective or approximately random procedures, which would invalidate any statistical test based on the data collected from such an experiment. In addition to assigning treatments at random, other possible sources of bias should be avoided by randomization: (a) The order in which measurements are done in the different experimental units should be randomized. Never measure all the experimental units receiving one level of treatment together and then all those receiving another level of treatment together. If the design includes blocks, then do the measurements by block, with the order of the different levels of treatment assigned at random within blocks. Be particularly careful if using more than one machine of the same type to make sure that the machines have been cross-calibrated. Likewise, if more than one person is working on an experiment, test that both people use the same criteria for taking measurements by comparing their readings when measuring a subsample of the same experimental units. (b) If several treatments are applied to each experimental unit in a sequential design, the order in which they are applied should be randomized. (c) Everything that can be randomized, should be randomized to avoid bias, and to ensure that the assumptions behind statistical tests of significance are fulfilled.

Randomization is of fundamental importance because: (a) It allows the organization of an experiment to be objective. (b) It prevents systematic errors from known and unknown sources of variation, because when randomized, these sources of variation should affect all treatments equally 'on average'. Randomization guarantees that the estimators obtained for the treatment effect and of the error variance are unbiased. This in turn allows valid conclusions to be derived from statistical tests.

5.6.3 Blocks and covariates

In designs with blocks we arrange the experimental units into homogeneous groups (according to some important characteristics). Each of these groups is called a *block*. Treatments are randomized within the blocks, normally

1. Replication → control of random variation → precision.

2. Randomization → elimination of systematic error → no bias.

3. Use of blocks → reduction of error variation caused by experimental unit heterogeneity → increased power of statistical tests.

Figure 5.1: The principles of experimental design and their 'purpose'.

with all treatments present in each block at least once. If blocking is successful it decreases the error variance because the systematic variation between the blocks can be separately accounted for in the analysis. If blocking is not successful, the estimator of error variance is not significantly reduced, but has fewer degrees of freedom[3].

Every effort should be made to use and retain a balanced design (i.e. blocks of the same size, and an equal number of replicates for all treatments) since this makes analysis and interpretation of results much simpler. There are sometimes advantages to the adoption of more complex (but still balanced) designs that use the same principles as blocking, but allocate treatments to groups based on more than one criterion. Examples of these are Latin square and Greco-Latin square designs. See Box 5.3.

Even if it is impossible to group the experimental units into homogeneous blocks, it may be possible to measure some relevant property of the experimental units before applying the treatments. Afterwards, this measured variable can be included as a covariate in an analysis of covariance (ANCOVA), or linear mixed effects (LME) or non-linear mixed effects (NLME) model. This may improve the performance of the statistical tests, by accounting for some of the random variability among experimental units. In any case, one should be careful with interpretation of ANCOVA results (Cochran, 1957; Smith, 1957). Including a covariate in a model does not lessen the requirement for proper randomization.

The flowchart in Figure 5.2 summarizes the designs most suitable for different situations based on the characteristics of the experimental units.

5.7 Experimental units and subsamples

It is very important to understand the difference between experimental units and subsamples. This is crucial because correct statistical analysis is only possible if we correctly identify the experimental units in our experiments. An *experimental unit* is the unit or 'thing' to which the treatment is assigned (at random) (e.g. a tray of plants). An experimental unit is not necessarily the unit that is measured, which can be smaller (e.g. a leaf from a treated plant). A measured object (= measurement unit) which is smaller than an experimental unit is called a subsample (or subunit).

In a simple design the experimental units are usually easy to identify. The same is true for non-hierarchical factorial designs (i.e. when the factors are not nested). However, in hierarchical designs, like split-plot designs, the experimental units for one factor may be nested within the larger experimental units used for another factor. In such a case, the error terms in an ANOVA will be different for the different factors, and will also have a different number of degrees of freedom. Another common situation occurs when treatments are applied to experimental units and the same units are measured or sampled repeatedly in time. In this case it is not appropriate to apply a factorial design, with time as one factor, in an ANOVA since this would assume that all observations are independent. Instead one should use a design that takes into account the correlation among the repeated measurements. See Box 5.4 for examples, and the flowchart in Figure 5.3 for a summary of how the relationship between experimental and measurement units affects the data analysis.

5.8 Pseudoreplication

Pseudoreplication involves the misidentification of the experimental unit as what is really the measurement unit, so giving a higher n than is truly the case. Examples of pseudoreplication are quite readily found in the scientific literature. In UV experiments pseudoreplication is particularly prevalent as researchers often mistake the plant for the lamp frame or filter as the experimental unit. Hurlbert (1984) was the first paper to bring this frequent problem of ecological research to light. Pseudoreplication is not just a small annoyance, it is a difficult

---

[3]It is not usually recommended to remove the blocks factor from a statistical analysis if the significance of this term in an ANOVA is non-significant. This post-hoc removal can only be justified when $F < 1.0$ for the blocks term.

Box 5.3: Experiment using a block design.

If the experimental units belong to two, or more, statistical populations, then the best option is to use a design that acknowledges this. Such populations could be distinct soil series, cultivars or even years.

For example if we have three different cultivars of a crop that are important in a certain region for which we want the conclusions from our experiment to be applicable, we need to include several experimental units for each of the cultivars. In this case, structure should be added by classifying the units into three blocks one for each cultivar. In this way the sensitivity of our test will not be affected by the additional variation due to cultivars, but the results will be applicable to the three of them, instead of just to one.

If we suspect that the cultivars may respond to UV differently, the appropriate design would be a factorial one, which would allow to assess this difference through the interaction term.

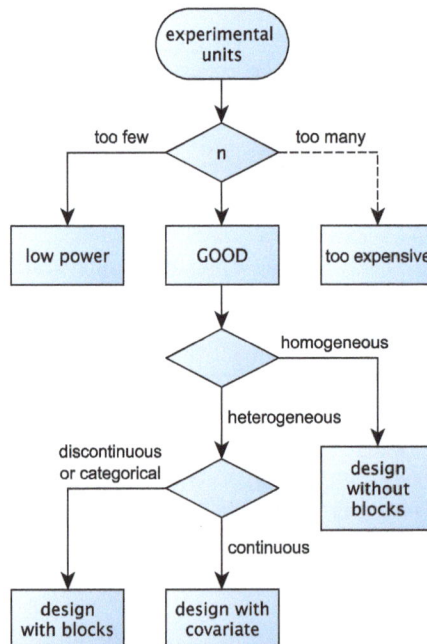

Figure 5.2: Flowchart describing the relationship between the characteristics of the experimental units, and the design of experiments used for statistical data analysis.

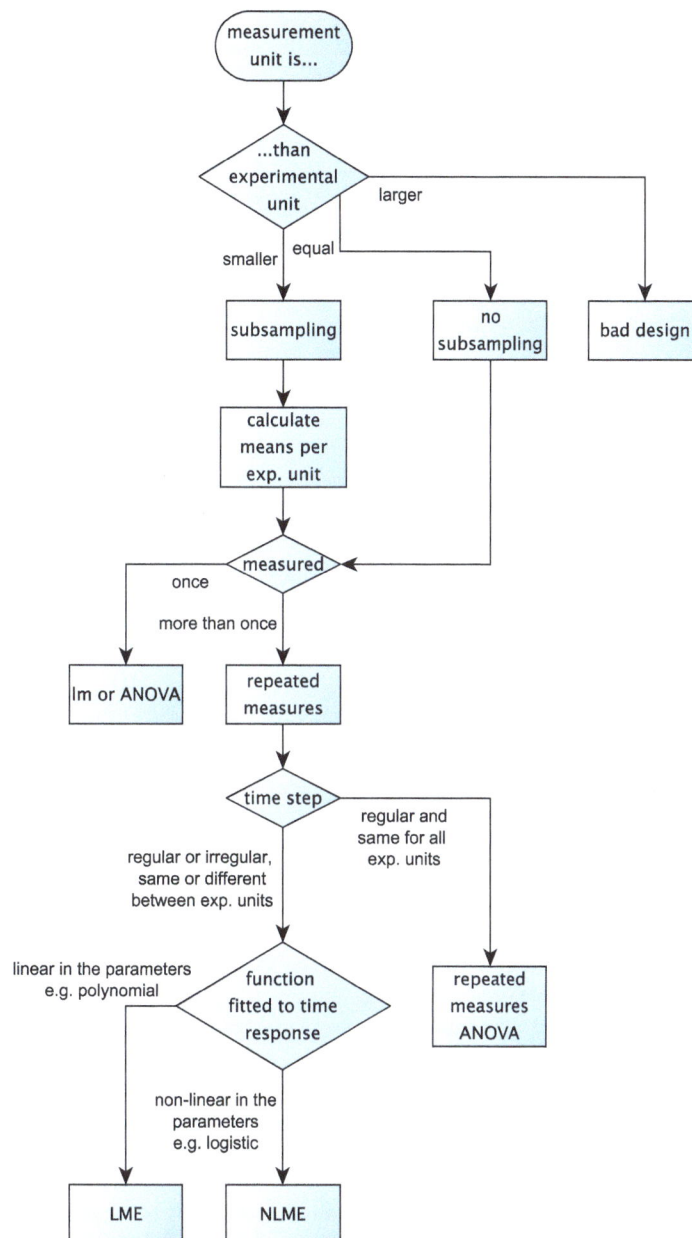

Figure 5.3: Flow chart describing the relationship between measurement units and experimental units, and its relationship to the design of experiments used for statistical data analysis. Although this flowchart only shows an option for analysing data from an experiment with subsampling for the calculation of means per experimental unit, other valid options also exist which allow the estimation of subsampling errors. ANOVA = analysis of variance, lm = linear model, LME = linear mixed effects model, NLME = non-linear mixed effects model.

Box 5.4: Experimental units and subsamples: example

1. In an experiment we grow three plants per pot. We have nine pots. The treatments are three different watering regimes, which are assigned to the pots. We measure photosynthesis on individual plants. We get three numbers per pot. The pots are the experimental units. The photosynthesis measurements from each plant are subsamples. The subsamples are not independent observations. The (random) assignment of the treatments was not done on the plants, so the plants are not experimental units. The treatments were assigned to the pots.

   $n = 3, N = 9$

   n is the number of true replicates. N is the number of experimental units in the whole experiment.

2. In an experiment we grow three plants per pot. We have nine pots. The treatments are three different foliar fertilizers, which are assigned (at random) to the plants within each pot. We measure photosynthesis on individual plants. We get three numbers per pot. The plants are the experimental units. The photosynthesis measurements from the plants are replicates. The three plants in each pot are not independent of each other. To account for this non-independence, the pots are treated as blocks in the statistical analysis. (Depending on the duration of the experiment and the differences in size between plants, in practice, using 27 pots with one plant per pot may be better.)

   $n = 9, N = 27$

3. In an experiment we grow one plant per pot. We have nine pots. The treatments are three different watering regimes, which are assigned (at random) to the pots. We measure photosynthesis on individual plants. We get one number per pot. The pots (and plants) are the experimental units. The measurement of each plant/pot is a replicate. The replicates are independent observations.

   $n = 3, N = 9$

---

problem in field experiments in ecology, and it frequently remains hidden within the Discussion and Conclusions sections of scientific papers. The most frequent scenario occurs when experiments are designed as a comparison of two *typical* experimental units but conclusions are applied to the whole population of units. In other words, conclusions are based on subsamples rather than on true replicates. Quinn and Keough (2002) describes this situation as a mismatch between the scale of treatments and the scale of replicates.

Pseudoreplication happens when subsamples are treated as replicates in the statistical analysis and interpretation.

When an experiment is thoughtfully planned and the researcher is certain about the identity the unit of replication, it should be possible to avoid pseudoreplication. However, where an oversight in planning or logistics has made pseudoreplication unavoidable, the additional assumptions involved in the interpretation of any statistical tests must be clearly indicated in reports or publications arising from the study. See Box 5.5 for examples of pseudoreplication and suggested ways of avoiding this problem.

5.9 Range of validity

The range of validity of the conclusions of an experiment derives from the population upon which the experiment was performed. These conclusions can not be extrapolated beyond the statistical population from which the experimental units were randomly chosen. The wider the range of validity, the more generally applicable the information obtained from an experiment will be. If we wish to perform an experiment with broad applicability we can increase the breadth of the sampled population, but in doing so we run the risk of having more heterogeneous experimental units. This can make treatment effects more difficult to detect, unless we take special measures to control this error variation.

If we do an experiment with only one inbred variety of a crop species the results and conclusions will apply only to that variety. If we mix the seeds from several varieties and assign the plants at random to the treatments, we enlarge the range of validity of the results and conclusions but we increase the random variation. We can control for this increase in variation by using several varieties as blocks. In this way, we increase the range of

Box 5.5: Pseudoreplication: examples

Problem 1: We want to study the effect of UV radiation on the growth of plants. We have two rooms, one with only normal lamps supplying PAR and one with some UV lamps in addition to the normal lamps. We randomly assign 20 plants to each room. We measure the height of the plants after one week. This experiment has 20 subsamples per treatment, but only one replicate. (The UV treatments were assigned at random to the rooms, not to the plants). It is not valid to treat the subsamples as replicates and try to draw conclusions about the effect of UV radiation, since we have only pseudoreplicates. In this case pseudoreplication adds the implicit **assumption** that the only difference between the rooms was the UV irradiance. Our statistical test really only answers the question: did the plants grow differently in the two rooms?

Solution 1: Several remedies are needed to obtain a wider range of validity from this experiment. Replicates in time can be created by repeating the experiment several times swapping the UV lamps between rooms. Such a situation is far from ideal because it is difficult to maintain plants and rooms in an equivalent state over time, since factors such as time of year and deterioration of lamps and filters must be accounted for. An additional step towards overcoming this problem would be to make a pre-experimental trial trying to create 'identical' conditions in both growth rooms and checking whether this results in any difference in plant performance.

Problem 2: We want to study the differences between high elevation and lowland meadow vegetation. We choose one typical lowland meadow (LM) and one high elevation meadow (HM). We establish 5 plots, located at random, within each meadow. In each plot we mark ten 1 m^2 sampling areas at random, and record the species composition within these areas.

Solution 2: It is not valid to try to answer the question 'is species x more frequent in lowland meadows than in highland meadows' using the plots as replicates, since we have only pseudoreplication. The context of the question posed is critical in this instance. For this question we have, one true replicate (one meadow of each type), 5 subsamples (the plots) within each meadow, and 10 subsubsamples within each subsample. If the experiment is performed as described, it can only answer the question: "Is species x more abundant in meadow 'LM' than meadow 'HM'?" If we want to draw conclusions about the two (statistical) populations of meadows, we should sample at random from those populations. For example compare five different highland meadows to five lowland meadows.

Problem 3: We have three UV-B lamp frames. The radiation from the lamps in one frame is filtered with cellulose di-acetate, in another frame the lamps are filtered with polyester, and in the third frame the lamps are not energized. We chose at random three groups of 100 seedlings and put one group under each frame. This is an un-replicated experiment, we have one experimental unit (one lamp frame) per treatment. If we use the plants as replicates, we commit pseudoreplication. The plants are subsamples, not replicates.

Solution 3: Ideally we would try to obtain more lamp frames. If this was impossible, a replicated experiment could be created by sub-dividing the area under each lamp frame into two separate parts, and allocating one half of each containing 50 seedlings to be filtered by cellulose di-acetate and the other half by polyester. The unenergised lamp treatment would have to be omitted, but this compromise would enable three replicates of the two treatments to be created. Each lamp frame would be a block.

Problem 4: We have nine UV-B lamp frames, three for each treatment. We put 100 plants under each frame as above. We do the statistical analysis based on the measurements on each plant considering the plants as replicates, we commit pseudoreplication. Our estimate of the error variance is wrong, as it does not reflect the variation among frames with the same treatment, and degrees of freedom are hugely inflated.

Solution 4: Alternatively, we do the experiment exactly as in the item above, but we calculate means per frame for the measured variable, and then use these means in the statistical analysis. In other words we analyse our data using the frames as replicates. We have a valid test of significance, because the error variance reflects the variation between equally treated experimental units (the lamp frames). Logistical constraints often require that a decision must be made about whether to measure more plants under fewer replicates, or fewer plants under more replicates. When such a compromise has to be made, it is nearly always preferable to have more experimental units and to measure fewer plants within each one, even at the expense of greater allocation of time and resources to the maintenance of filters and lamp frames.

147

validity of the results and conclusions without increasing the random variation.

The design of an experiment involves many compromises concerning its range of validity, these relate to trade-offs between generality, precision, and realism, not to mention cost.

5.9.1 Factorial experiments

One way of increasing the range of validity of the results of a scientific study is to use a factorial design for an experiment. For example, if we study the effects of UV radiation on well-watered plants only, then the conclusions of our study will be valid only for well-watered plants. If we include a second factor in our design with three levels, 'drought', 'mild drought' and 'well watered' and include all the treatment combinations possible, e.g. based on the three levels of watering and three levels of UV attenuation we have nine treatments and the range of validity is greatly expanded. In addition we can statistically test whether these two factors interact[4]. Factorial experiments are a very powerful and useful design but if too many factors and levels are included their interpretation may become difficult. Factorial experiments with many factors and/or levels result in a statistical analysis based on many different contrasts[5]. In this case, we should remember that in most analyzes we wish to obtain *P*-values per contrast, rather than per experiment. If we want to keep the experiment-wise risk level constant, we should use a tight contrast-wise risk level, or adjust the *P*-values.

The flowchart in Figure 5.4 describes the relationships between the structure of treatments and data analysis.

5.10 When not to make multiple comparisons

In the biological literature, multiple comparison procedures such as Tukey's HSD are frequently used in situations where other tests would be more effective and easier to interpret. In some other cases multiple comparisons are used in situations in which they give misleading results. We will discuss the two most common cases of misuse of these tests.

5.10.1 Dose response curves

If we have several levels of the same treatment, for example several different irradiances of UV, the most power-ful (capable of reliably detecting effects) test is to fit a regression (either linear or non-linear) of the response on the "dose". From such an analysis we get information on the relationship between dose and response, whether it is increasing or decreasing, linear or curvilinear, even its shape can be inferred. In many cases we can also calculate a confidence band around the fitted function. This is all useful and interpretable information.

If instead we calculate for example HSD or LSD and compare the responses to all possible pairs of doses, we discard the information about the ordering of the doses, and we can get results that do not address our research hypothesis, for example, concluding that all pairs of adjacent doses do not differ significantly while the extreme doses do differ significantly. In this type of experiments we are really interested in the slope and shape of a dose response curve.

Multiple comparisons should be carried out only when there is no ordering in the levels of a treatment of factor. A regression or ANCOVA model should be used instead, if a reasonable model can be fitted.

5.10.2 Factorial experiments

Factorial experiments are very useful and as discussed above allow extending the range of validity of our conclusions. However, the main advantage of factorial experiments is that the significance of interactions can be tested. The most important part of the statistical analysis of a factorial experiment is an ANOVA with a model including main effects and interactions. Only after this test, one can *in addition*, test the differences among different pairs of treatments. This *a posteriori* test provides little extra information and should be done only if the interaction term is statistically significant. Otherwise, if the levels of a factor are discrete and unordered, and its main effect is significant in ANOVA, a multiple comparison test can be used to compare the levels of a factor (but not the individual combinations of levels of the different factors).

5.11 Presenting data in figures

When using hierarchical designs such as split-plots, the experimental errors relevant to comparisons between levels of the different factors will be different. In such designs, it is especially important to indicate which error term the error bars in figures show. In all cases, the figure and table legends should state which statistics are depicted by the error bars, and also the number of replicates

---

[4]The main difference between a design with blocks and a factorial design, is that in the first case it is assumed that there is no interaction between treatments and blocks, and the interaction term is used as an estimator of the error variance while in a factorial design we estimate the error variance based on equally treated units and we do not need to assume the lack of interaction effect.

[5]In ANOVA, a contrast is a combination of factor level means (two or more) based on multipliers that add to zero. Each contrast defines one of many possible tests of significance.

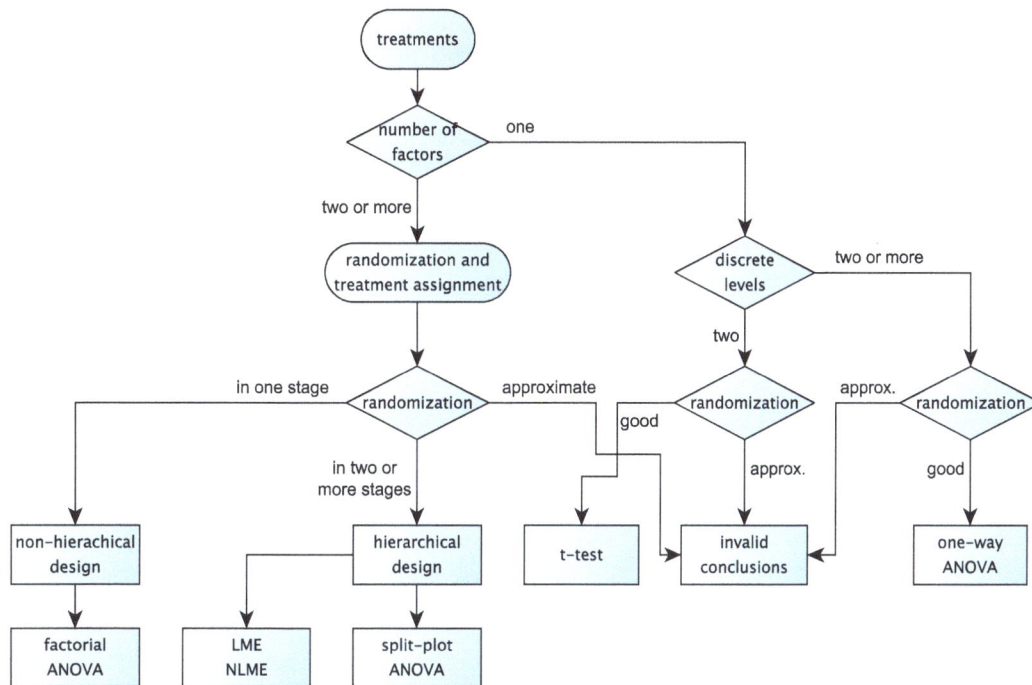

Figure 5.4: Flow chart describing the relationship between the structure of the treatments and the design of experiments used for statistical data analysis. ANOVA = analysis of variance, LME = linear mixed effects model, NLME = non-linear mixed effects model.

involved in their calculation.

If what one wants to describe is the variability in the original sampled population then one should use the standard deviation (s.d.), whose expected value does not depend on the number of replicates. If one wants to compare treatment or group means, one should use confidence intervals (c.i.) or the standard error of the mean (s.e.). These last two statistics decrease in size with an increase in the number of replicates, reflecting increased reliability of the estimated means.

5.12 | Recommendations

In this section we list recommendations related to the statistical design of experiments for studying the effects of UV radiation on plants. See sections 2.6.1 and 2.6.2 for recommendations on manipulation of UV radiation, section 3.15 for recommendations on how to quantify UV radiation, and section 4.9 for recommendations about plant growing conditions.

1. Avoid pseudoreplication. In most experiments where UV irradiance is manipulated either with lamps or filters, individual plants or pots are not

true replicates. The true replicates are the lamp- or filter frames. Design your experiments and analyze your data taking this into account.

2. Avoid pseudoreplication. In experiments under controlled conditions, if treatments are assigned to chambers or rooms, these chambers or rooms are the true replicates, not the plants within them. To have true replicates you will need several rooms per treatment. When this is not possible, the experiment can be repeated in time, using the same chambers or rooms, but reassigning the treatments, so that each chamber is used for a different treatment in each iteration of the experiment.

3. Include enough true replicates in your experiment. Many effects of UV radiation are small in relation to the mean and subtle in comparison to natural levels of variability. Reliably detecting these effects requires well-replicated experiments.

As a rough rule of thumb, aim at having at least 32 plants per treatment for growth, ecophysiology and metabolite measurements, in at least four true replicates composed each of eight measured plants in controlled environment and greenhouse studies. When making gas-exchange readings, it is usually impractical to measure from more than four to six plants in each of the four true replicates. If you aim to make an ecologically-orientated outdoor study it would be advisable to have at least five true replicates. In addition to issues of replication, in most cases you will need to have unmeasured border plants surrounding your experimental plants so as to avoid edge effects. Alternatively the positions of plants can be rotated every few days.

4. Preferably use a design with blocks to control all known or expected sources of error (or background variation) that would otherwise confound the treatment effects. Examples of such sources of variation are different benches in a greenhouse, or measurements done by different observers. If you suspect that a known source of error may interact with your UV treatment, you may decide to consider it as a covariate in your model.

5. Within the blocks, use randomization to neutralize the effects of other error sources. Randomization is of fundamental importance and should be done properly: i.e. a device that generates a truly random assignment of treatments to experimental units should be employed. Such devices can be as simple as a coin, or die, or can be a table of random numbers in a book, or (pseudo-) random numbers generated by a computer or a pocket calculator.

6. When designing an experiment use power analysis to determine the number of replicates needed, unless you have a long experience of performing similar experiments, and a good existing understanding of the amount of variability you expect to find in your results.

7. If your experiment yields a non-significant difference, use *a posteriori* power analysis to demonstrate, if possible, that biologically important differences would have been detected had they existed. However, if power analysis reveals that your experiment was unable to detect biologically important differences, you should repeat the experiment with more replicates or an improved design.

5.13 Further reading

A reliable on-line resource is the *Statistics Stack Exchange* (`http://stats.stackexchange.com/`).

There are many text books on statistics, but most introductory statistical texts do not discuss the type of hierarchical models usually needed for UV experiments. We recommend the classic book 'Planning of Experiments' by Cox (1958) as a good introduction to the principles behind the design of experiments. 'Experimental Design and Data Analysis for Biologists' by Quinn and Keough (2002) describes both experimental design and analysis of data at an intermediate level. The books 'Ecological Models and Data in R' (Bolker, 2008) and 'Mixed effects models and extensions in ecology with R' (Zuur, Ieno, Walker et al., 2009) give advanced up-to-date accounts of statistical methods, and approaches, including a discussion of hypothesis testing and model fitting. While 'A Beginner's Guide to R' (Zuur, Ieno and Meesters, 2009) and 'Introductory Statistics with R' (Dalgaard, 2002) are gentle introductions to data analysis with R.

Bibliography

Albert, A. and P. Gege (2006). 'Inversion of irradiance and remote sensing reflectance in shallow water between 400 and 800 nm for calculations of water and bottom properties'. In: *Applied Optics* 45, pp. 2331-2343. DOI: 10.1364/AO.45.002331 (cit. on p. 20).

Albert, A. and C. D. Mobley (2003). 'An analytical model for subsurface irradiance and remote sensing reflectance in deep and shallow case-2 waters'. In: *Optics Express* 11.22, pp. 2873-2890. URL: http://www.opticsexpress.org/abstract.cfm?URI=OPEX-11-22-2873 (cit. on pp. 15, 20).

Allen, L. H., H. W. Gausman and W. A. Allen (1975). 'Solar Ultraviolet Radiation in Terrestrial Plant Communities'. In: *Journal of Environmental Quality* 4.3, pp. 285-294. DOI: 10.2134/jeq1975.0047242500040030001x (cit. on p. 11).

Anderson, J., W. Chow and D. Goodchild (1988). 'Thylakoid Membrane Organisation in Sun/Shade Acclimation'. In: *Functional Plant Biology* 15, pp. 11-26. DOI: 10.1071/PP9880011 (cit. on p. 124).

Aphalo, P. J. (2003). 'Do current levels of UV-B radiation affect vegetation? The importance of long-term experiments'. In: *New Phytologist* 160. Invited commentary, pp. 273-276. DOI: 10.1046/j.1469-8137.2003.00905.x (cit. on p. 2).

Aphalo, P. J., R. Tegelberg and R. Julkunen-Tiitto (1999). 'The modulated UV-B irradiation system at the University of Joensuu'. In: *Biotronics* 28, pp. 109-120. URL: http://ci.nii.ac.jp/naid/110006175827/en (cit. on pp. 39, 48-50).

Arends, G., R. K. A. M. Mallant, E. van Wensveen and J. M. Gouman (1988). *A fog chamber for the study of chemical reactions*. Tech. rep. Report - ECN - 210. Petten, Netherlands (cit. on p. 126).

Arfsten, D. P., D. J. Schaeffer and D. C. Mulveny (1996). 'The effects of near ultraviolet radiation on the toxic effects of polycyclic aromatic hydrocarbons in animals and plants: A review'. In: *Ecotoxicology and Environment Safety* 33, pp. 1-24. DOI: 10.1006/eesa.1996.0001 (cit. on p. 19).

Arts, M. T., R. D. Robarts, F. Kasai, M. J. Waiser, V. P. Tumber, A. J. Plante, H. Rai and H. J. de Lange (2000). 'The attenuation of ultraviolet radiation in high dissolved organic carbon waters of wetlands and lakes on the northern Great Plains'. In: *Limnology and Oceano-*

graphy 45, pp. 292-299. URL: http://www.jstor.org/stable/2670483 (cit. on p. 18).

Austin, A. T. and C. L. Ballaré (2010). 'Dual role of lignin in plant litter decomposition in terrestrial ecosystems'. In: *Procceedings of the National Academy of Sciences of the U.S.A* 107, pp. 4618-4622. DOI: 10.1073/pnas.0909396107 (cit. on p. 2).

Bakker, J. C., G. P. A. Bot, H. Challa and N. J. Vand de Braak, eds. (1995). *Greenhouse climate control: An integrated approach*. Wageningen, The Netherlands: Wageningen Academic Publishers. 279 pp. ISBN: 978-90-74134-17-0. DOI: 10.3920/978-90-8686-501-7 (cit. on p. 138).

Ballaré, C. L., M. M. Caldwell, S. D. Flint, S. A. Robinson and J. F. Bornman (2011). 'Effects of solar ultraviolet radiation on terrestrial ecosystems. Patterns, mechanisms, and interactions with climate change'. In: *Photochemical and Photobiological Sciences* 10 (2), pp. 226-241. DOI: 10.1039/C0PP90035D (cit. on p. 1).

Barnes, P. W., S. D. Flint, J. R. Slusser, W. Gao and R. J. Ryel (2008). 'Diurnal changes in epidermal UV transmittance of plants in naturally high UV environments'. In: *Physiologia Plantarum* 133, pp. 363-372. DOI: 10.1111/j.1399-3054.2008.01084.x (cit. on p. 21).

Bazzaz, F. A. and R. W. Carlson (1982). 'Photosynthetic acclimation to variability in the light environment of early and late successional plants'. In: *Oecologia* 54, pp. 313-316. DOI: 10.1007/BF00379999 (cit. on p. 124).

Belzile, C., S. Demers, G. A. Ferreyra, I. Schloss, C. Nozais, K. Lacoste, B. Mostajir, S. Roy, M. Gosselin, E. Pelletier et al. (2006). 'UV effects on marine planktonic food webs: A synthesis of results from mesocosm studies'. In: *Photochemistry and Photobiology* 82, pp. 850-856. DOI: 10.1562/2005-09-27-RA-699 (cit. on p. 136).

Belzile, C., S. Demers, D. R. S. Lean, B. Mostajir, S. Roy, S. de Mora, D. Bird, M. Gosselin, J. P. Chanut and M. Levasseur (1998). 'An experimental tool to study the effects of ultraviolet radiation on planktonic communities: A mesocosm approach'. In: *Environmental Technology* 19, pp. 667-682. DOI: 10.1080/09593331908616723 (cit. on p. 136).

Bentham (1997). *A Guide to Spectroradiometry: Instruments & Applications for the Ultraviolet.* Tech. rep. Reading, U.K.: Bentham Instruments Ltd (cit. on p. 110).

Bérces, A., A. Fekete, S. Gáspár, P. Gróf, P. Rettberg, G. Horneck and G. Rontó (1999). 'Biological UV dosimeters in the assessment of the biological hazard from environmental radiation'. In: *Journal of Photochemistry and Photobiology, B* 53.1-3, pp. 36-43. DOI: 10.1016/S1011-1344(99)00123-2 (cit. on p. 78).

Berger, D. S. (1976). 'The sunburning ultraviolet meter: design and performance'. In: *Photochemistry and Photobiology* 24, pp. 587-593. DOI: 10.1111/j.1751-1097.1976.tb06877.x (cit. on p. 80).

Beytes, C., ed. (2003). *Ball Red Book: Greenhouses and equipment.* 17th ed. Batavia, IL, USA: Ball Publishing. 272 pp. ISBN: 1883052343 (cit. on p. 138).

Bickford, E. D. and S. Dunn (1972). *Lighting for plant growth.* Ohio, USA: Kent State University Press. x + 221. ISBN: 0873381165 (cit. on p. 48).

Bilger, W., T. Johnsen and U. Schreiber (2001). 'UV-excited chlorophyll fluorescence as a tool for the assessment of UV-protection by epidermins of plants'. In: *Journal of Experimental Botany* 52, pp. 2007-2014. DOI: 10.1093/jexbot/52.363.2007 (cit. on pp. 21, 23).

Bilger, W., M. Veit, L. Schreiber and U. Schreiber (1997). 'Measurement of leaf epidermal transmittance of UV radiation by chlorophyll fluorescence'. In: *Physiologia Plantarum* 101.4, pp. 754-763. DOI: 10.1111/j.1399-3054.1997.tb01060.x (cit. on p. 21).

Bird, R. E. and C. Riordan (1986). 'Simple solar spectral model for direct and diffuse irradiance on horizontal and tilted planes at the Earth's surface for cloudless atmospheres'. English. In: *Journal of Climate and Applied Meteorology* 25.1, pp. 87-97. ISSN: 0733-3021. DOI: 10.1175/1520-0450(1986)025%3C0087:SSSMFD%3E2.0.CO;2 (cit. on p. 97).

Bischof, K., I. Gómez, M. Molis, D. Hanelt, U. Karsten, U. Lüder, M. Y. Roleda, K. Zacher and C. Wiencke (2006). 'Ultraviolet radiation shapes seaweed communities'. In: *Reviews in Evironmental Science and Biotechnology* 51, pp. 141-166. DOI: 10.1007/s11157-006-0002-3 (cit. on p. 131).

Björn, L. O. (1995). 'Estimation of fluence rate from irradiance measurements with a cosine-corrected sensor'. In: *Journal of Photochemistry and Photobiology B Biology* 29, pp. 179-183. DOI: 10.1016/1011-1344(95)07135-O (cit. on p. 74).

Björn, L. O., ed. (2007). *Photobiology: The Science of Life and Light.* 2nd ed. Springer. 684 pp. ISBN: 0387726543 (cit. on pp. 2, 27, 76).

Björn, L. O. and T. M. Murphy (1985). 'Computer calculation of solar ultraviolet radiation at ground level'. In: *Physiologie Végétale* 23, pp. 555-561 (cit. on pp. 97, 98).

Björn, L. O. and A. H. Teramura (1993). 'Simulation of daylight ultraviolet radiation and effects of ozone depletion'. In: *Environmental UV Photobiology.* Ed. by A. R. Young, L. O. Björn, J. Moan and W. Nultsch. New York: Plenum Press, pp. 41-71. ISBN: 0-306-44443-7 (cit. on pp. 38, 40, 41, 67, 111).

Björn, L. O. and T. C. Vogelmann (1996). 'Quantifying light and ultraviolet radiation in plant biology'. In: *Photochemistry and Photobiology* 64, pp. 403-406. DOI: 10.1111/j.1751-1097.1996.tb03084.x (cit. on pp. 75, 76).

Bloom, A. A., J. Lee-Taylor, S. Madronich, D. J. Messenger, P. I. Palmer, D. S. Reay and A. R. McLeod (2010). 'Global methane emission estimates from ultraviolet irradiation of terrestrial plant foliage'. In: *New Phytologist* 187.2, pp. 417-425. DOI: 10.1111/j.1469-8137.2010.03259.x (cit. on p. 2).

Bolker, B. M. (2008). *Ecological Models and Data in R.* Princeton University Press. 408 pp. ISBN: 0691125228 (cit. on p. 150).

Booker, F. L., E. L. Fiscus, R. B. Philbeck, A. S. Heagle, J. E. Miller and W. W. Heck (1992). 'A supplemental ultraviolet-B radiation system for open-top chambers'. In: *Journal of Environmental Quality* 21, pp. 56-61 (cit. on p. 127).

Bornman, J. F. and T. C. Vogelmann (1988). 'Penetration by blue and UV radiation measured by fiber optics in spruce and fir needles'. In: *Physiologia Plantarum* 72.4, pp. 699-705. DOI: 10.1111/j.1399-3054.1988.tb06368.x (cit. on p. 21).

Bowman, W. D. and J. N. Demas (1976). 'Ferrioxalate actinometry - Warning on its correct use'. In: *Journal of Physical Chemistry* 80.21, pp. 2434-2435. ISSN: 0022-3654. DOI: 10.1021/j100562a025 (cit. on p. 77).

Braslavsky, S. E. (2007). 'Glossary of terms used in Photochemistry 3(rd) Edition (IUPAC Recommendations 2006)'. In: *Pure and Applied Chemistry* 79.3, pp. 293-465. DOI: 10.1351/pac200779030293 (cit. on p. 71).

Bricaud, A., M. Babin, A. Morel and H. Claustre (1995). 'Variability in the chlorophyll-specific absorption coefficients of natural phytoplankton: analysis and parameterization'. In: *Journal of Geophysical Research* 100.C7, pp. 13321-13332 (cit. on p. 17).

Bricaud, A., A. Morel and L. Prieur (1981). 'Absorption by dissolved organic matter of the sea (yellow substance) in the UV and visible domains'. In: *Limnology and Oceanography* 26.1, pp. 43-53. URL: http://www.jstor.org/stable/2835805 (cit. on p. 16).

Brown, M. J., G. G. Parker and N. E. Posner (1994). 'A survey of ultraviolet-B radiation in forests'. In: *Journal of Ecology* 82.4, pp. 843–854. URL: `http://www.jstor.org/stable/2261448` (cit. on pp. 11, 14).

Buiteveld, H., J. H. M. Hakvoort and M. Donze (1994). 'The optical properties of pure water'. In: *Proceedings of SPIE "Ocean Optics XII"*. Vol. 2258. International Society for Optical Engineering, pp. 174–183 (cit. on p. 16).

Cabello-Pasini, A., E. Aguirre-von-Wobeser and F. L. Figueroa (2000). 'Photoinhibition of photosynthesis in *Macrocystis pyrifera* (*Phaeophyceae*), *Chondrus crispus* (*Rhodophyceae*) and *Ulva lactuca* (*Chlorophyceae*) in outdoor culture systems'. In: *Journal of Photochemistry and Photobiology B: Biology* 57, pp. 169–178. DOI: `10.1016/S1011-1344(00)00095-6` (cit. on p. 134).

Caldwell, M. M. (1971). 'Solar UV irradiation and the growth and development of higher plants'. In: *Photophysiology*. Ed. by A. C. Giese. Vol. 6. New York: Academic Press, pp. 131–177. ISBN: 012282606X (cit. on pp. xxiii, 1, 25, 105, 111).

Caldwell, M. M. and S. D. Flint (1994a). 'Lighting considerations in controlled environments for nonphotosynthetic plant responses to blue and ultraviolet radiation'. In: *Proceedings of the International Lighting in Controlled Environments Workshop*. Vol. NASA-CP-95-3309, pp. 113–124 (cit. on p. 50).

– (1994b). 'Stratospheric ozone reduction, solar UV-B radiation and terrestrial ecosystems'. In: *Climatic Change* 28.4, pp. 375–394. DOI: `10.1007/BF01104080` (cit. on p. 1).

Caldwell, M. M. and S. D. Flint (2006). 'Use and Evaluation of Biological Spectral UV Weighting Functions for the Ozone Reduction Issue'. In: *Environmental UV Radiation: Impact on Ecosystems and Human Health and Predictive Models*. Ed. by F. Ghetti, G. Checcucci and J. F. Bornman. Vol. 57. NATO Science Series. Proceedings of the NATO Advanced Study Institute on Environmental UV Radiation: Impact on Ecosystems and Human Health and Predictive Models Pisa, Italy June 2001. Dordrecht: Springer, pp. 71–84. ISBN: 978-1-4020-3695-8. DOI: `10.1007/1-4020-3697-3` (cit. on p. 102).

Caldwell, M. M., S. D. Flint and P. S. Searles (1994). 'Spectral balance and UV-B sensitivity of soybean: A field experiment'. In: *Plant, Cell and Environment* 17.3, pp. 267–276. DOI: `10.1111/j.1365-3040.1994.tb00292.x` (cit. on p. 1).

Caldwell, M. M., W. G. Gold, G. Harris and C. W. Ashurst (1983). 'A modulated lamp system for solar UV-B (280-320 nm) supplementation studies in the field'. In: *Photochemistry and Photobiology* 37, pp. 479–485. DOI: `10.1111/j.1751-1097.1983.tb04503.x` (cit. on pp. 48, 136).

Caldwell, M. M., A. H. Teramura and M. Tevini (1989). 'The Changing Solar Ultraviolet Climate and the Ecological Consequences for Higher Plants'. In: *Trends in Ecology & Evolution* 4.12, pp. 363–367. DOI: `10.1016/0169-5347(89)90100-6` (cit. on p. 1).

Campbell, G. S. and J. M. Norman (1998). *An Introduction to Environmental Biophysics*. 2nd ed. New York: Springer. 286 pp. ISBN: 0-387-94937-2 (cit. on p. 15).

Cen, Y.-P. and J. F. Bornman (1993). 'The effect of exposure to enhanced UV-B radiation on the penetration of monochromatic and polychromatic UV-B radiation in leaves of Brassica napus'. In: *Physiologia Plantarum* 87.3, pp. 249–255. DOI: `10.1111/j.1399-3054.1993.tb01727.x` (cit. on p. 21).

Christie, J. M. (2007). 'Phototropin Blue-Light Receptors'. In: *Annual Review of Plant Biology* 58, pp. 21–45. DOI: `10.1146/annurev.arplant.58.032806.103951` (cit. on p. 1).

Christie, J. M., A. S. Arvai, K. J. Baxter, M. Heilmann, A. J. Pratt, A. O'Hara, S. M. Kelly, M. Hothorn, B. O. Smith, K. Hitomi et al. (2012). 'Plant UVR8 Photoreceptor Senses UV-B by Tryptophan-Mediated Disruption of Cross-Dimer Salt Bridges'. In: *Science*. DOI: `10.1126/science.1218091` (cit. on p. 1).

Cochran, W. G. (1957). 'Analysis of covariance its nature and uses'. In: *Biometrics* 13, pp. 261–281. URL: `http://www.jstor.org/stable/2527916` (cit. on p. 143).

Cohen, J. (1977). *Statistical Power Analysis for the Behavioral Sciences*. Revised edition. New York: Academic Press. 474 pp. (cit. on p. 139).

Coleman, A., R. Sarkany and S. Walker (2008). 'Clinical ultraviolet dosimetry with a CCD monochromator array spectroradiometer'. In: *Physics in Medicine and Biology* 53.18, pp. 5239–5255. DOI: `10.1088/0031-9155/53/18/026` (cit. on p. 94).

Coohill, T. P. (Oct. 1992). 'Action spectroscopy and stratospheric ozone depletion'. In: *UV-B monitoring workshop: a review of the science and status of measuring and monitoring programs*. Science and Policy Associates, Washington D.C., pp. C89–C112 (cit. on p. 24).

Cooley, N. M., H. M. F. Truscott, M. G. Holmes and T. H. Attridge (2000). 'Outdoor ultraviolet polychromatic action spectra for growth responses of *Bellis perennis* and *Cynosurus cristatus*'. In: *Journal of Photochemistry and Photobiology B: Biology* 59.1-3, pp. 64–71. DOI: `10.1016/S1011-1344(00)00141-X` (cit. on p. 27).

Cox, D. R. (1958). *Planning of Experiments*. New York: John Wiley & Sons. 308 pp. (cit. on pp. 142, 150).

Cox, D. R. and N. Reid (2000). *The Theory of the Design of Experiments*. 1st ed. Chapman and Hall/CRC. 314 pp. ISBN: 158488195X (cit. on p. 140).

Crawley, M. J. (2005). *Statistics: An Introduction using R.* Wiley. 342 pp. ISBN: 0470022981 (cit. on p. 116).

– (2007). *The R Book.* John Wiley and Sons Ltd. 950 pp. ISBN: 0470510242 (cit. on p. 116).

Cullen, J. J. and P. J. Neale (1997). 'Biological weighting functions for describing the effects of ultraviolet radiation on aquatic systems'. In: *The effects of ozone depletion on aquatic ecosystems.* Ed. by D.-P. Häder. Academic Press. Chap. 6, pp. 97–118. ISBN: 0123991730 (cit. on pp. 23, 26).

Dalgaard, P. (2002). *Introductory Statistics with R.* Statistics and Computing. New York: Springer. xv + 267. ISBN: 0 387 95475 9 (cit. on pp. 116, 150).

D'Antoni, H. L., L. J. Rothschild, C. Schultz, S. Burgess and J. W. Skiles (2007). 'Extreme environments in the forests of Ushuaia, Argentina'. In: *Geophysical Research Letters* 34.22. ISSN: 0094-8276. DOI: 10.1029/2007GL031096 (cit. on p. 92).

D'Antoni, H. L., L. J. Rothschild and J. W. Skiles (2008). 'Reply to comment by Stephan D. Flint et al. on "Extreme environments in the forests of Ushuaia, Argentina"'. In: *Geophysical Research Letters* 35.13. ISSN: 0094-8276. DOI: 10.1029/2008GL033836 (cit. on p. 92).

de la Rosa, T. M., R. Julkunen-Tiitto, T. Lehto and P. J. Aphalo (2001). 'Secondary metabolites and nutrient concentrations in silver birch seedlings under five levels of daily UV-B exposure and two relative nutrient addition rates'. In: *New Phytologist* 150, pp. 121–131. DOI: 10.1046/j.1469-8137.2001.00079.x (cit. on p. 60).

Deckmyn, G., E. Cayenberghs and R. Ceulemans (2001). 'UV-B and PAR in single and mixed canopies grown under different UV-B exclusions in the field'. In: *Plant Ecology* 154, 125–133. DOI: 10.1023/A:1012920716047 (cit. on p. 11).

Dekker, A. G. (1993). 'Detection of optical water quality parameters for eutrophic waters by high resolution remote sensing'. PhD thesis. Vrije Universiteit Amsterdam (cit. on pp. 15, 17, 18, 20).

DeLucia, E. H., T. A. Day and T. C. Vogelman (1992). 'Ultraviolet-B and visible light penetration into needles of two species of subalpine conifers during foliar development'. In: *Plant, Cell and Environment* 15.8, pp. 921–929. DOI: 10.1111/j.1365-3040.1992.tb01024.x (cit. on p. 21).

Demas, J. N., W. D. Bowman, E. F. Zalewski and R. A. Velapoldi (1981). 'Determination of the quantum yield of the ferrioxalate actinometer with electrically calibrated radiometers'. In: *Journal of Physical Chemistry* 85.19, pp. 2766–2771. ISSN: 0022-3654. DOI: 10.1021/j150619a015 (cit. on p. 78).

den Outer, P., H. Slaper, J. Kaurola, A. Lindfors, A. Kazantzidis, A. Bais, U. Feister, J. Junk, M. Janouch and W. Josefsson (2010). 'Reconstructing of erythemal ultraviolet radiation levels in Europe for the past 4 decades'. In: *Journal of Geophysical Research (Atmospheres)* 115.D14, D10102. DOI: 10.1029/2009JD012827 (cit. on pp. 97, 98).

Díaz, S., C. Camilión, J. Escobar, G. Deferrari, S. Roy, K. Lacoste, S. Demers, C. Belzile, G. Ferreyra, S. Gianesella et al. (2006). 'Simulation of ozone depletion using ambient irradiance supplemented with UV lamps.' In: *Photochemistry and Photobiology* 82.4, pp. 857–864. DOI: 10.1562/2005-09-28-RA-700 (cit. on pp. 48, 136).

Díaz, S., C. Camilión, K. Lacoste, J. Escobar, S. Demers, S. Gianesella and S. Roy (2003). 'Simulation of increasing UV radiation as a consequence of ozone depletion'. In: *Proceedings SPIE's 48th annual meeting: Ultraviolet ground- and spacebased measurements, models and effects III.* Ed. by J. Slusser, J. Herman and W. Gao. SPIE: The International Sociaty for Optical Engineering, pp. 216–227 (cit. on p. 136).

Díaz, S. B., J. E. Frederick, T. Lucas, C. R. Booth and I. Smolskaia (1996). 'Solar ultraviolet irradiance at Tierra del Fuego: Comparison of measurements and calculations over a full annual cycle'. In: *Geophysical Research Letters* 23.4, pp. 355–358. DOI: 10.1029/96GL00253 (cit. on p. 10).

Díaz, S. B., J. H. Morrow and C. R. Booth (2000). 'UV physics and optics'. In: *The Effects of UV Radiation in the Marine Environment.* Ed. by S. de Mora, S. Demers and M. Vernet. Cambridge University Press, pp. 35–71 (cit. on p. 95).

Diffey, B. L. (1987). 'A comparison of dosimeters used for solar ultraviolet radiometry'. In: *Photochem Photobiol* 46, pp. 55–60. DOI: 10.1111/j.1751-1097.1987.tb04735.x (cit. on p. 79).

– (1989). *Radiation Measurement in Photobiology.* London: Academic Press. 230 pp. ISBN: 0122158407 (cit. on p. 111).

Dixon, J. M., M. Taniguchi and J. S. Lindsey (2005). 'PhotochemCAD 2: a refined program with accompanying spectral data bases for photochemical calculations'. In: *Photochemistry and Photobiology* 81.1, pp. 212–213. DOI: 10.1111/j.1751-1097.2005.tb01544.x (cit. on p. 17).

Döhring, T., M. Köfferlein, S. Thiel and H. K. Seidlitz (1996). 'Spectral shaping of artificial UV-B irradiation for vegetation stress research'. In: *Journal of Plant Physiology* 148, pp. 115–119. DOI: 10.1016/S0176-1617(96)80302-6 (cit. on pp. 50, 54).

Donahue, W. F., D. W. Schindler, S. J. Page and M. P. Stainton (1998). 'Acid-induced changes in DOC quality in an experimental whole-lake manipulation'. In: *Environmental Science and Technology* 32, pp. 2954-2960. DOI: 10.1021/es980306u (cit. on p. 18).

Du, H., R.-. C. A. Fuh, J. Li, L. A. Corkan and J. S. Lindsey (1998). 'PhotochemCAD: a computer-aided design and research tool in photochemistry'. In: *Photochemistry and Photobiology* 68.2, pp. 141-142. DOI: 10.1111/j.1751-1097.1998.tb02480.x (cit. on p. 17).

Dunne, R. P. (1999). 'Polysulphone film as an underwater dosimeter for solar ultraviolet-B radiation in tropical latitudes'. In: *Marine Ecology Progress Series* 189, pp. 53-63. DOI: 10.3354/meps189053 (cit. on p. 78).

Eichler, H.-. J., A. Fleischner, J. Kross, M. Krystek, H. Lang, H. Niedrig, H. Rauch, G. Schmahl, H. Schoenebeck, E. Sedlmayr et al. (1993). *Bergmann, Schaefer: Lehrbuch der Experimentalphysik Band 3: Optik*. Ed. by H. Niedrig. Verlag Walter de Gruyter Berlin/New York (cit. on p. 3).

Einstein, A. (1910). 'Theorie der Opaleszenz von homogenen Flüssigkeiten und Flüssigkeitsgemischen in der Nähe des kritischen Zustandes'. In: *Annalen der Physik IV. Folge* 33.16, pp. 1275-1298 (cit. on p. 16).

Eisinger, W., T. E. Swartz, R. A. Bogomolni and L. Taiz (2000). 'The ultraviolet action spectrum for stomatal opening in broad bean'. In: *Plant Physiology* 122, pp. 99-105. DOI: http://dx.doi.org/10.1104/pp.122.1.99 (cit. on p. 131).

Engelsen, O. and A. Kylling (2005). 'Fast simulation tool for ultraviolet radiation at the earth's surface'. In: *Optical Engineering* 44.4. DOI: 10.1117/1.1885472 (cit. on pp. 97, 98).

Fernández, E., J. M. Figuera and A. Tobar (1979). 'Use of the potassium ferrioxalate actinometer below 254-nm'. In: *Journal of Photochemistry* 11.1, pp. 69-71. ISSN: 0047-2670. DOI: 10.1016/0047-2670(79)85008-X (cit. on pp. 77, 78).

Figueroa, F. L., A. Bueno, N. Korbee, R. Santos, L. Mata and A. Schuenhoff (2008). 'Accumulation of mycosporine-like amino acids in *Asparagopsis armata* grown in tanks with fishpond effluents of gilthead sea bream, *Asparus aurata*'. In: *Journal of the World Aquaculture Society* 39, pp. 692-699. DOI: 10.1111/j.1749-7345.2008.00199.x (cit. on p. 134).

Figueroa, F. L., S. Salles, J. Aguilera, C. Jiménez, J. Mercado, B. Viñegla, A. Flores-Moya and M. Altamirano (1997). 'Effects of solar radiation on photoinhibition and pigmentation in the red alga *Porphyra leucosticta*'. In: *Marine Ecology Progress Series* 151, pp. 81-90. DOI: 10.3354/meps151081 (cit. on p. 62).

Figueroa, F. L., R. Santos, R. Conde-Álvarez, L. Mata, J. L. Gómez-Pinchetti, J. Matos, P. Huovinen, A. Schüehoff and J. Silva (2006). 'The use of chlorophyll fluorescence for monitoring photosynthetic conditions of two tank cultivated red macroalgae using fishpond effluents'. In: *Botanica Marina* 49, pp. 275-282. DOI: 10.1515/BOT.2006.035 (cit. on p. 134).

Fioletov, V., L. McArthur, J. Kerr and D. Wardle (2001). 'Long-term variations of UV-B irradiance over Canada estimated from Brewer observations and derived from ozone and pyranometer measurements'. In: *Journal of Geophysical Research* 106.D19, pp. 23009-23027. DOI: 10.1029/2001JD000367 (cit. on p. 97).

Fioletov, V. E., M. G. Kimlin, N. Krotkov, L. J. B. McArthur, J. B. Kerr, D. I. Wardle, J. R. Herman, R. Meltzer, T. W. Mathews and J. Kaurola (2004). 'UV index climatology over the United States and Canada from ground-based and satellite estimates'. In: *Journal of Geophysical Research - Atmospheres* 109.D22. ISSN: 0148-0227. DOI: 10.1029/2004JD004820 (cit. on p. 97).

Flenley, J. R. (1992). 'Ultraviolet-B insolation and the altitudinal forest limit'. In: *Nature and dynamics of forest savanna boundaries*. Ed. by P. A. Furley, J. Proctor and J. A. Ratter. London: Chapman & Hall, pp. 273-282 (cit. on p. 9).

Flint, S. D., C. L. Ballare, M. M. Caldwell and R. L. McKenzie (2008). 'Comment on "Extreme environments in the forests of Ushuaia, Argentina" by Hector D'Antoni et al.' In: *Geophysical Research Letters* 35.13. ISSN: 0094-8276. DOI: 10.1029/2008GL033570 (cit. on p. 92).

Flint, S. D. and M. M. Caldwell (1996). 'Scaling plant ultraviolet spectral responses from laboratory action spectra to field spectral weighting factors'. In: *Journal of Plant Physiology* 148, pp. 107-114. DOI: 10.1016/S0176-1617(96)80301-4 (cit. on p. 25).

Flint, S. D. and M. M. Caldwell (1998). 'Solar UV-B and visible radiation in tropical forest gaps: measurements partitioning direct and diffuse radiation'. In: *Global Change Biology* 4.8, pp. 863-870. DOI: 10.1046/j.1365-2486.1998.00191.x (cit. on p. 14).

Flint, S. D. and M. M. Caldwell (2003). 'A biological spectral weighting function for ozone depletion research with higher plants'. In: *Physiologia Plantarum* 117, pp. 137-144. DOI: 10.1034/j.1399-3054.2003.1170117.x (cit. on pp. 25, 105, 112, 113).

Flint, S. D., R. J. Ryel, T. J. Hudelson and M. M. Caldwell (2009). 'Serious complications in experiments in which UV doses are effected by using different lamp heights'. In: *Journal of Photochemistry and Photobiology, B* 97.1, pp. 48-53. DOI: 10.1016/j.jphotobiol.2009.07.010 (cit. on p. 38).

Franklin, L. A. and R. M. Forster (1997). 'The changing irradiance environment: Consequences for marine macrophyte physiology, productivity and ecology'. In: *European Journal of Phycology* 32, pp. 207-232. DOI: 10.1080/09670269710001737149 (cit. on p. 62).

Frederick, J. E., P. F. Soulen, S. B. Diaz, I. Smolskaia, C. R. Booth, T. Lucas and D. Neuschuler (1993). 'Solar Ultraviolet Irradiance Observed From Southern Argentina: September 1990 to March 1991'. In: *Journal of Geophysical Research* 98, pp. 8891-8897. DOI: 10.1029/93JD00030 (cit. on p. 10).

Frigaard, N.-. U., K. L. Larsen and R. P. Cox (1996). 'Spectrochromatography of photosynthetic pigments as a fingerprinting technique for microbial phototrophs'. In: *FEMS Microbiology Ecology* 20, pp. 69-77. DOI: 10.1111/j.1574-6941.1996.tb00306.x (cit. on pp. 17, 19).

Fröhlich, C. and J. Lean (2004). 'Solar radiative output and its variability: evidence and mechanisms'. In: *The Astronomy and Astrophysics Review* 12, pp. 273-320. DOI: 10.1007/s00159-004-0024-1 (cit. on p. 8).

Furusawa, Y., L. E. Quintern, H. Holtschmidt, P. Koepke and M. Saito (1998). 'Determination of erythema-effective solar radiation in Japan and Germany with a spore monolayer film optimized for the detection of UVB and UVA-results of a field campaign'. In: *Appl Microbiol Biotechnol* 50, pp. 597-603. DOI: 10.1007/s002530051341 (cit. on p. 78).

García-Pichel, F. (1995). 'A scalar irradiance fiber-optic microprobe for the measurement of ultraviolet radiation at high spatial resolution'. In: *Photochemistry and Photobiology* 61, pp. 248-254. DOI: 10.1111/j.1751-1097.1995.tb03967.x (cit. on p. 21).

Gege, P. (1998). 'Characterization of the phytoplankton in Lake Constance for classification by remote sensing'. In: *Archiv für Hydrobiologie, Special issues: Advances in Limnology* 53, pp. 179-193 (cit. on p. 16).

- (2004). 'The water color simulator WASI: an integrating software tool for analysis and simulation of optical in situ spectra'. In: *Computers and Geosciences* 30, pp. 523-532. DOI: 10.1016/j.cageo.2004.03.005 (cit. on p. 20).

Geiss, O. (2003). *Manual for polysulphone dosimeter.* Tech. rep. EUR 20981 EN. European Union. URL: http://publications.jrc.ec.europa.eu/repository/bitstream/111111111/1227/1/EUR%2020981%20EN.pdf (cit. on pp. 78-80).

Ghetti, F., H. Herrmann, D.-. P. Häder and H. K. Seidlitz (1999). 'Spectral dependence of the inhibition of photosynthesis under simulated global radiation in the unicellular green alga *Dunaliella salina*'. In: *Journal of Photochemistry and Photobiology B: Biology* 48, pp. 166-173. DOI: 10.1016/S1011-1344(99)00043-3 (cit. on p. 26).

Goldstein, S. and J. Rabani (2008). 'The ferrioxalate and iodide-iodate actinometers in the UV region'. In: *Journal of Photochemistry and Photobiology A-Chemistry* 193.1,

pp. 50-55. ISSN: 1010-6030. DOI: 10.1016/j.jphotochem.2007.06.006 (cit. on pp. 76, 77).

Gómez, I., F. L. Figueroa, P. Huovinen, N. Ulloa and V. Morales (2005). 'Photosynthesis of the red alga *Gracilaria chilensis* under natural solar radiation in an estuary in southern Chile'. In: *Aquaculture* 244, pp. 369-382. DOI: h10.1016/j.aquaculture.2004.11.037 (cit. on p. 62).

Gómez, I., F. L. Figueroa, N. Ulloa, V. Morales, C. Lovengreen, P. Huovinen and S. Hess (2004). 'Photosynthesis in 18 intertidal macroalgae from Southern Chile'. In: *Marine Ecology Progress Series* 270, pp. 103-116. DOI: 10.3354/meps270103 (cit. on p. 62).

Gordon, H. R. and A. Y. Morel (1983). *Remote assessment of ocean color for interpretation of satellite visible imagery: a review.* Ed. by R. T. Barber, C. N. K. Mooers, M. J. Bowman and B. Zeitzschel. Vol. 4. Lecture Notes on Coastal and Estuarine Studies. New York: Springer Verlag (cit. on p. 17).

Gorton, H. L. (2010). 'Biological action spectra'. In: *Photobiological Sciences Online.* Ed. by K. C. Smith. American Society for Photobiology. URL: http://www.photobiology.info/Gorton.html (cit. on p. 24).

Götz, M., A. Albert, S. Stich, W. Heller, H. Scherb, A. Krins, C. Langebartels, H. K. Seidlitz and D. Ernst (2010). 'PAR modulation of the UV-dependent levels of flavonoid metabolites in *Arabidopsis thaliana* (L.) Heynh. leaf rosettes: cumulative effects after a whole vegetative growth period'. In: *Protoplasma* 243, pp. 95-103. DOI: 10.1007/s00709-009-0064-5 (cit. on pp. 26, 27).

Goulas, Y., Z. G. Cerovic, A. Cartelat and I. Moya (2004). 'Dualex: a new instrument for field measurements of epidermal ultraviolet absorbance by chlorophyll fluorescence'. In: *Applied Optics* 43.23, pp. 4488-4496. DOI: 10.1364/AO.43.004488 (cit. on p. 21).

Gould, K. S., T. C. Vogelmann, T. Han and M. J. Clearwater (2002). 'Profiles of photosynthesis within red and green leaves of *Quintinia serrata*'. In: *Physiologia Plantarum* 116.1, pp. 127-133. DOI: 10.1034/j.1399-3054.2002.1160116.x (cit. on p. 21).

Graedel, T. E. and P. J. Crutzen (1993). *Atmospheric Change: An Earth System Perspective.* New York: WH Freeman. 446 pp. ISBN: board 0-7167-2334-4, paper 0-7167-2332-8 (cit. on pp. 10, 27).

Grant, R. H. (1998). 'Ultraviolet irradiance of inclined planes at the top of plant canopies'. In: *Agricultural and Forest Meteorology* 89, pp. 281-293. DOI: 10.1016/S0168-1923(97)00067-1 (cit. on p. 11).

- (1999a). 'Potential effect of soybean heliotropism on ultraviolet-B irradiance and dose'. In: *Agronomy*

Journal 91, pp. 1017-1023. DOI: `doi:10.2134/agronj1999.9161017x` (cit. on p. 11).

- (1999b). 'Ultraviolet-B and photosynthetically active radiation environment of inclined leaf surfaces in a maize canopy and implications for modeling'. In: *Agricultural and Forest Meteorology* 95, pp. 187-201. DOI: `10.1016/S0168-1923(99)00023-4` (cit. on p. 11).

- (2004). 'UV Radiation Penetration in Plant Canopies'. In: *Encyclopedia of Plant and Crop Science*, pp. 1261-1264. DOI: `10.1081/E-EPCS-120010624` (cit. on p. 11).

Green, A. E. S. and J. H. Miller (1975). 'Measures of biologically active radiation in the 280-340 nm region. Impacts of climate change on the environment'. In: CIAP Monograph 5, Part 1. Chap. 2.2.4 (cit. on p. 111).

Green, A. E. S., T. Sawada and E. P. Shettle (1974). 'The middle ultraviolet reaching the ground'. In: *Photochemistry and Photobiology* 19, pp. 251-259. DOI: `10.1111/j.1751-1097.1974.tb06508.x` (cit. on pp. xxiii, 25, 111, 113).

Grifoni, D., F. Sabatini, G. Zipoli and M. Viti (2009). 'Action spectra affect variability in the climatology of biologically effective UV radiation (UVBE)'. In: *Poster presentation at the Final Seminar of COST Action 726, 13-14 May 2009, Warsaw, Poland* (cit. on p. 105).

Grifoni, D., G. Zipoli, M. Viti and F. Sabatini (2008). 'Latitudinal and seasonal distribution of biologically effective UV radiation affecting human health and plant growth'. In: *Proceedings of 18th International Congress of Biometeorology, 22-26 September 2008, Tokyo, Japan* (cit. on p. 105).

Haan, H. D. (1972). 'Molecule-size distribution of soluble humic compounds from different natural waters'. In: *Freshwater Biology* 2, pp. 235-241. DOI: `10.1111/j.1365-2427.1972.tb00052.x` (cit. on p. 16).

- (1993). 'Solar UV-light penetration and photodegradation of humic substances in peaty lake water'. In: *Limnology and Oceanográphy* 38, pp. 1072-1076. URL: `http://www.jstor.org/stable/2838095` (cit. on p. 16).

Haan, H. D., D. Boer and T (1987). 'Applicability of light absorbance and fluorescence as measures of concentration and molecular size of dissolved organic carbon in humic lake Tjeukemeer'. In: *Water Research* 21, pp. 731-734. DOI: `10.1016/0043-1354(87)90086-8` (cit. on p. 16).

Häder, D.-P. and F. L. Figueroa (1997). 'Photoecophysiology of marine macroalgae'. In: *Photochemistry and Photobiology* 66, pp. 1-14. DOI: `10.1111/j.1751-1097.1997.tb03132.x` (cit. on p. 131).

Häder, D.-P., E. W. Helbling, C. E. Williamson and R. C. Worrest (2011). 'Effects of UV radiation on aquatic eco-

systems and interactions with climate change'. In: *Photochemical and Photobiological Sciences* 10 (2), pp. 242-260. DOI: `10.1039/C0PP90036B` (cit. on p. 1).

Häder, D.-P., M. Lebert, A. Flores, C. Jiménez, J. Mercado, S. Salles, J. Aguilera and F. L. Figueroa (1996). 'Photosynthetic oxygen production and PAM fluorescence in the brown alga *Padina pavonica* (Linnaeus) Lamouroux measured in the field under solar radiation'. In: *Marine Biology* 127, pp. 61-66. DOI: `10.1007/BF00993644` (cit. on p. 62).

Häder, D.-P., M. Lebert, R. Marangoni and G. Colombetti (1999). 'ELDONET—European light dosimeter network hardware and software'. In: *Journal of Photochemistry and Photobiology B: Biology* 52, pp. 51-58. DOI: `10.1016/S1011-1344(99)00102-5` (cit. on p. 95).

Häder, D.-P., M. Lebert, M. Schuster, L. del Ciampo, E. W. Helbling and R. McKenzie (2007). 'ELDONET—a decade of monitoring solar radiation on five continents'. In: *Photochem Photobiol* 83, pp. 1348-1357. DOI: `10.1111/j.1751-1097.2007.00168.x` (cit. on pp. 14, 15).

Hakvoort, J. H. M. (1994). 'Absorption of light by surface water'. PhD thesis. Delft University of Technology (cit. on pp. 16, 17).

Hannay, J. W. and D. J. Millar (1986). 'Phytotoxicity of phthalate plasticisers. I. Diagnosis and commercial implications'. In: *Journal of Experimental Botany* 37, pp. 883-897. DOI: `10.1093/jxb/37.6.883` (cit. on p. 129).

Hardwick, R. C. and R. A. Cole (1987). 'Plastics that kill plants'. In: *Outlook on Agriculture* 16.13, pp. 100-104 (cit. on p. 129).

Hargreaves, B. R. (2003). 'Water column optics and penetration of UVR'. In: *UV effects in aquatic organisms and ecosystems*. Ed. by E. W. Helbling and H. Zagarese. Cambridge, UK: The Royal Society of Chemistry, pp. 59-105. ISBN: 0854043012 (cit. on pp. 15, 19).

Hatchard, C. G. and C. A. Parker (1956). 'A new sensitive chemical actinometer .2. Potassium ferrioxalate as a standard chemical actinometer'. In: *Proceedings of the Royal Society of London Series A-Mathematical and Physical Sciences* 235.1203, pp. 518-536. DOI: `10.1098/rspa.1956.0102` (cit. on pp. 76, 78).

Hegglin, M. I. and T. G. Shepherd (2009). 'Large climate-induced changes in ultraviolet index and stratosphere-to-troposphere ozone flux'. In: *Nature Geoscience* advance online publication, pp. 687-691. DOI: `10.1038/ngeo604` (cit. on p. 10).

Heijde, M. and R. Ulm (2012). 'UV-B photoreceptor-mediated signalling in plants'. In: *Trends in Plant Science*. DOI: `10.1016/j.tplants.2012.01.007` (cit. on p. 1).

Hessen, D. O. and E. V. Donk (1994). 'Effects of UV-radiation of humic water on primary and secondary production'. In: *Water, Air & Soil Pollution* 75, pp. 325–338. DOI: 10.1007/BF00482944 (cit. on p. 19).

Hessen, D. O. and L. J. Tranvik (1998). *Aquatic humic substances*. Berlin Heidelberg: Springer-Verlag. 361 pp. ISBN: 3540639101 (cit. on p. 18).

Hirose, T. (2005). 'Development of the Monsi-Saeki Theory on Canopy Structure and Function'. In: *Annals of Botany* 95, pp. 483–494. DOI: 10.1093/aob/mci047 (cit. on p. 11).

Hogewoning, S. W., P. Douwstra, G. Trouwborst, W. van Ieperen and J. Harbinson (2010). 'An artificial solar spectrum substantially alters plant development compared with usual climate room irradiance spectra'. In: *Journal of Experimental Botany* 61.5, pp. 1267–1276. DOI: 10.1093/jxb/erq005 (cit. on p. 127).

Holmes, M. G. (1984). 'Light Sources'. In: *Techniques in Photomorphogenesis*. Ed. by H. Smith and M. G. Holmes. Academic press, pp. 43–79. ISBN: 0126529906 (cit. on p. 42).

– (1997). 'Action spectra for UV-B effects on plants: monochromatic and polychromatic approaches for analysing plant responses'. In: *Plants and UV-B - responses to environmental change*. Ed. by P. J. Lumsden. Cambridge University Press, pp. 31–50. ISBN: 0521572223 (cit. on p. 24).

Holmes, M. G. and D. R. Keiller (2002). 'Effects of pubescence and waxes on the reflectance of leaves in the ultraviolet and photosynthetic wavebands: a comparison of a range of species'. In: *Plant Cell and Environment* 25.1, pp. 85–93. DOI: 10.1046/j.1365-3040.2002.00779.x (cit. on p. 21).

Horneck, G., P. Rettberg, E. Rabbow, W. Strauch, G. Seckmeyer, R. Facius, G. Reitz, K. Strauch and J. U. Schott (1996). 'Biological dosimetry of solar radiation for different simulated ozone column thicknesses'. In: *Journal of Photochemistry and Photobiology B-biology* 32.3, pp. 189–196. ISSN: 1011-1344. DOI: 10.1016/1011-1344(95)07219-5 (cit. on p. 78).

Hulst, H. C. van de (1981). *Light scattering by small particles*. unabridged and corrected republication of the work originally published in 1957 by John Wiley & Sons Inc. New York. New York: Dover Publications Inc. (cit. on p. 16).

Hunt, J. E. (1997). 'Ultraviolet-B radiation and its effects on New Zealand trees'. Ph.D. Dissertation. Canterbury, New Zealand: Lincoln University, p. 106 (cit. on p. 55).

Hunt, J. E. and D. L. McNeil (1998). 'Nitrogen status affects UV-B sensitivity of cucumber'. In: *Australian Journal of Plant Physiology* 25.1, pp. 79–86. DOI: 10.1071/PP97102 (cit. on p. 48).

Huovinen, P. and I. Gómez (2011). 'Spectral attenuation of solar radiation in Patagonian fjords and coastal waters and implications for algal photobiology'. In: *Continental Shelf Research* 31, pp. 254–259. DOI: 10.1016/j.csr.2010.09.004 (cit. on pp. 19, 62).

Huovinen, P. S., H. Penttilä and M. R. Soimasuo (2003). 'Spectral attenuation of solar ultraviolet radiation in humic lakes in Central Finland'. In: *Chemosphere* 51, pp. 205–214. DOI: 10.1016/S0045-6535(02)00634-3 (cit. on pp. 15, 18).

Hurlbert, S. H. (1984). 'Pseudoreplication and the design of ecological field experiments'. In: *Ecological Monographs* 54.2, pp. 187–211. DOI: 10.2307/1942661 (cit. on p. 143).

Ibdah, M., A. Krins, H. K. Seidlitz, W. Heller, D. Strack and T. Vogt (2002). 'Spectral dependence of flavonol and betacyanin accumulation in *Mesembryanthemum crystallinum* under enhanced ultraviolet radiation'. In: *Plant, Cell and Environment* 25.9, pp. 1145–1154. DOI: doi:10.1046/j.1365-3040.2002.00895.x (cit. on pp. 25–27).

Iqbal, M. (1983). *An introduction to solar radiation*. Academic Press Canada (cit. on pp. 3, 8).

Jagger, J. (1967). *Introduction to research in ultraviolet photobiology*. Englewood Cliffs, NJ, USA: Prentice-Hall. 164 pp. ISBN: 0134955722 (cit. on p. 76).

Jansen, M. A. K. and J. F. Bornman (2012). 'UV-B radiation: from generic stressor to specific regulator'. In: *Physiologia Plantarum* 145.4, pp. 501–504. ISSN: 1399-3054. DOI: 10.1111/j.1399-3054.2012.01656.x (cit. on p. 2).

Jenkins, G. I. (2009). 'Signal transduction in responses to UV-B radiation'. In: *Annual Review of Plant Biology* 60, pp. 407–431. DOI: 10.1146/annurev.arplant.59.032607.092953 (cit. on p. 1).

Jerlov, N. (1976). *Marine optics*. 2nd. Amsterdam: Elsevier. 246 pp. ISBN: 0444414908 (cit. on pp. 19, 20).

Jones, H. G. (1992). *Plants and Microclimate: A Quantitative Approach to Environmental Plant Physiology*. 2nd ed. Cambridge University Press. 456 pp. ISBN: 0521425247 (cit. on p. 138).

Jones, L. W. and B. Kok (1966). 'Photoinhibition of Chloroplast Reactions. II. Multiple Effects'. In: *Plant Physiology* 41, pp. 1044–1049. DOI: 10.1104/pp.41.6.1044 (cit. on p. 25).

Julkunen-Tiitto, R., H. Häggman, P. J. Aphalo, A. Lavola, R. Tegelberg and T. Veteli (2005). 'Growth and defense in deciduous trees and shrubs under UV-B'. In: *Environmental Pollution* 137, pp. 404–414. DOI: 10.1016/j.envpol.2005.01.050 (cit. on p. 1).

Kalbin, G., S. Li, H. Olsman, M. Pettersson, M. Engwall and Å. Strid (2005). 'Effects of UV-B in biological and chemical systems: equipment for wavelength dependence determination'. In: *Journal of Biochemical and Biophysical Methods* 65, pp. 1-12. DOI: 10.1016/j.jbbm.2005.09.001 (cit. on p. 42).

Kalbina, I., S. Li, G. Kalbin, L. Björn and Å. Strid (2008). 'Two separate UV-B radiation wavelength regions control expression of different molecular markers in *Arabidopsis thaliana*'. In: *Functional Plant Biology* 35.3, pp. 222-227. DOI: 10.1071/FP07197 (cit. on p. 42).

Kalle, K. (1966). 'The problem of the gelbstoff in the sea'. In: *Oceanography and Marine Biology Annual Review* 4, pp. 91-104 (cit. on p. 16).

Karabourniotis, G. and J. F. Bornman (1999). 'Penetration of UV-A, UV-B and blue light through the leaf trichome layers of two xeromorphic plants, olive and oak, measured by optical fibre microprobes'. In: *Physiologia Plantarum* 105, pp. 655-661. DOI: 10.1034/j.1399-3054.1999.105409.x (cit. on p. 21).

Karentz, D., J. E. Cleaver and D. L. Mitchell (1991). 'Cell survival characteristics and molecular responses of Antarctic phytoplankton to ultraviolet-B radiation'. In: *Journal of Phycology* 27, pp. 326-341. DOI: 10.1111/j.0022-3646.1991.00326.x (cit. on p. 19).

Keiller, D. R., S. A. H. Mackerness and M. G. Holmes (2003). 'The action of a range of supplementary ultraviolet (UV) wavelengths on photosynthesis in Brassica napus L. in the natural environment: effects on PSII, CO2 assimilation and level of chloroplast proteins'. In: *Photosynthesis Research* 75.2, pp. 139-150. DOI: 10.1023/A:1022812229445 (cit. on p. 27).

Khanh, T. Q. and W. Dähn (1988). 'Die Ulbrichtsche Kugel. Theorie und Anwendungsbeispiele in der optischen Strahlungsmeßtechnik'. In: *Photonik*, pp. 6-9 (cit. on p. 95).

Kirk, A. D. and C. Namasivayam (1983). 'Errors in ferrioxalate actinometry'. In: *Analytical Chemistry* 55.14, pp. 2428-2429. ISSN: 0003-2700. DOI: 10.1021/ac00264a053 (cit. on p. 77).

Kirk, J. T. O. (1980). 'Spectral absorption properties of natural waters: Contribution of the soluble and particulate fractions to light absorption in some inland waters of southeastern Australia'. In: *Australian Journal of Marine and Freshwater Research* 31, pp. 287-296. DOI: 10.1071/MF9800287 (cit. on p. 19).

– (1991). 'Volume scattering function, average cosine, and the underwater light field'. In: *Limnology and Oceanography* 36.3, pp. 455-467. URL: http://www.jstor.org/stable/2837511 (cit. on p. 20).

– (1994a). *Light and photosynthesis in aquatic ecosystems.* 2nd ed. Cambridge, UK: Cambridge University Press. 509 pp. ISBN: 0 521 45353 4 (cit. on pp. 15, 16, 18-20, 94, 95).

– (1994b). 'Optics of UV-B radiation in natural waters'. In: *Arch. Hydrobiol. Beih. Ergebn. Limnol* 43, pp. 1-166 (cit. on pp. 15, 18).

Kjeldstad, B., B, O. Frette, S. R. Erga, H. I. Browman, P. Kuhn, R. Davis, W. Miller and J. J. Stamnes (2003). 'UV (280 to 400 nm) optical properties in a Norwegian fjord system and an intercomparison of underwater radiometers'. In: *Marine Ecology Progress Series* 256, pp. 1-11. DOI: 10.3354/meps256001 (cit. on p. 95).

Koepke, P., H. D. Backer, A. Bais, A. Curylo, K. Eerme, U. Feister, B. Johnsen, J. Junk, A. Kazantzidis, J. Krzyscin et al. (2006). 'Modelling solar UV radiation in the past: comparison of algorithms and input data'. In: *Remote Sensing of Clouds and the Atmosphere XI.* Ed. by J. R. Slusser, K. Schäfer and A. Comeron. Vol. 6362. Proceedings of SPIE. DOI: 10.1117/12.687682 (cit. on p. 97).

Kolb, C. A., U. Schreiber, R. Gademann and E. E. Pfündel (2005). 'UV-A screening in plants determined using a new portable fluorimeter'. In: *Photosynthetica* 43.3, pp. 371-377. DOI: 10.1007/s11099-005-0061-7 (cit. on p. 21).

Kopp, G. and J. L. Lean (2011). 'A new, lower value of total solar irradiance: Evidence and climate significance'. In: *Geophys. Res. Lett.* 38.1, pp. L01706-. DOI: 10.1029/2010GL045777 (cit. on p. 8).

Kotilainen, T., A. Lindfors, R. Tegelberg and P. J. Aphalo (2011). 'How realistically does outdoor UV-B supplementation with lamps reflect ozone depletion: An assessment of enhancement errors'. In: *Photochemistry and Photobiology* 87, pp. 174-183. DOI: 10.1111/j.1751-1097.2010.00843.x (cit. on pp. 12, 25, 85, 102-104, 109).

Kotilainen, T., R. Tegelberg, R. Julkunen-Tiitto, A. Lindfors and P. J. Aphalo (2008). 'Metabolite specific effects of solar UV-A and UV-B on alder and birch leaf phenolics'. In: *Global Change Biology* 14, pp. 1294-1304. DOI: 10.1111/j.1365-2486.2008.01569.x (cit. on p. 55).

Kotilainen, T., T. Venäläinen, R. Tegelberg, A. Lindfors, R. Julkunen-Tiitto, S. Sutinen, R. B. O'Hara and P. J. Aphalo (2009). 'Assessment of UV Biological Spectral Weighting Functions for Phenolic Metabolites and Growth Responses in Silver Birch Seedlings'. In: *Photochemistry and Photobiology* 85, pp. 1346-1355. DOI: 10.1111/j.1751-1097.2009.00597.x (cit. on p. 55).

Kowalczuk, P., M. Zabłocka, S. Sagan and K. Kuliński (2010). 'Fluorescence measured in situ as a proxy of CDOM absorption and DOC concentration in the Baltic Sea'. In: *Oceanologia* 52.3, pp. 431-471 (cit. on pp. 18, 19).

Kreuter, A. and M. Blumthaler (2009). 'Stray light correction for solar measurements using array spectrometers'. In: *Review of Scientific Instruments* 80.9, 096108, p. 096108. DOI: 10.1063/1.3233897 (cit. on p. 94).

Krizek, D. T. and R. M. Mirecki (2004). 'Evidence for phytotoxic effects of cellulose acetate in UV exclusion studies'. In: *Environmental and Experimental Botany* 51, pp. 33-43. DOI: 10.1016/S0098-8472(03)00058-3 (cit. on p. 55).

Kuhn, H. J., S. E. Braslavsky and R. Schmidt (1989). 'Chemical actinometry'. In: *Pure and Applied Chemistry* 61.2, pp. 187-210. ISSN: 0033-4545. DOI: 10.1351/pac198961020187 (cit. on p. 76).

Kujanpää, J., N. Kalakoski and T. Koskela (2010). 'Three Years of O3M SAF Surface UV Products'. In: *Proceedings of the 2010 EUMETSAT Meteorological Satellite Conference, Cordoba, Spain* (cit. on pp. 98, 99).

Kylling, A., A. R. Webb, A. F. Bais, M. Blumthaler, R. Schmitt, S. Thiel, A. Kazantzidis, R. Kift, M. Misslbeck, B. Schallhart et al. (2003). 'Actinic flux determination from measurements of irradiance'. English. In: *Journal of Geophysical Research* 108.D16. ISSN: 0148-0227. DOI: 10.1029/2002JD003236 (cit. on p. 99).

Langhans, R. W. and T. W. Tibbitts, eds. (1997). *Plant growth chamber handbook*. Vol. SR-99. North Central Regional Research Publication 340. Iowa Agriculture and Home Economics Experiment Station. URL: http://www.controlledenvironments.org/Growth_Chamber_Handbook/Plant_Growth_Chamber_Handbook.htm (cit. on p. 48).

Lean, D. (1998). 'Attenuation of solar radiation in humic waters'. In: *Aquatic humic substances*. Ed. by D. O. Hessen and L. J. Tranvik. Berlin Heidelberg: Springer-Verlag, pp. 109-124 (cit. on p. 18).

Lee, J. and H. H. Seliger (1964). 'Quantum yield of ferrioxalate actinometer'. In: *Journal of Chemical Physics* 40.2, pp. 519-523. ISSN: 0021-9606. DOI: 10.1063/1.1725147 (cit. on pp. 76, 78).

Lee, Z. P., K. L. Carder and R. A. Arnone (2002). 'Deriving inherent optical properties from water color: a multiband quasi-analytical algorithm for optically deep water'. In: *Applied Optics* 41.27, pp. 5755-5772. DOI: 10.1364/AO.41.005755 (cit. on p. 20).

Lester, R. A., A. V. Parisi, M. G. Kimlin and J. Sabburg (2003). 'Optical properties of poly(2,6-dimethyl-1,4-phenylene oxide) film and its potential for a long-term solar ultraviolet dosimeter'. In: *Physics in Medicine and Biology* 48.22, pp. 3685-3698. DOI: 10.1088/0031-9155/48/22/005 (cit. on p. 79).

Leszczynski, K. (2002). 'Advances in Traceability of Solar Ultraviolet Radiation Measurements'. PhD thesis. University of Helsinki (cit. on p. 82).

Liboriussen, L., F. Landkildehus, M. Meerhoff, M. E. Bramm, M. Søndergaard, K. Christoffersen, K. Richardson, M. Søndergaard, T. L. Lauridsen and E. Jeppesen (2005). 'Global warming: Design of a flow-through shallow lake mesocosm climate experiment'. In: *Limnology and Oceanography: Methods* 3, pp. 1-9. DOI: 10.4319/lom.2005.3.1 (cit. on p. 136).

Lindfors, A., A. Heikkilä, J. Kaurola, T. Koskela and K. Lakkala (2009). 'Reconstruction of Solar Spectral Surface UV Irradiances Using Radiative Transfer Simulations'. In: *Photochemistry and Photobiology* 85.5, pp. 1233-1239. ISSN: 0031-8655. DOI: 10.1111/j.1751-1097.2009.00578.x (cit. on pp. 99, 100).

Lindfors, A., J. Kaurola, A. Arola, T. Koskela, K. Lakkala, W. Josefsson, J. A. Olseth and B. Johnsen (2007). 'A method for reconstruction of past UV radiation based on radiative transfer modeling: Applied to four stations in northern Europe'. In: *Journal of Geophysical Research* 112.D23, D23201. DOI: 10.1029/2007JD008454 (cit. on p. 99).

Lindfors, A., A. Tanskanen, A. Arola, R. van der A, A. Bais, U. Feister, M. Janouch, W. Josefsson, T. Koskela, K. Lakkala et al. (2009). 'The PROMOTE UV Record: Toward a Global Satellite-Based Climatology of Surface Ultraviolet Irradiance'. In: *IEEE Journal of Selected Topics in Applied Earth Observation and Remote Sensing* 2.3, pp. 207-212. ISSN: 1939-1404. DOI: 10.1109/JSTARS.2009.2030876 (cit. on p. 99).

Long, S. P. and J.-E. Hällgren (1987). 'Measurement of CO_2 assimilation by plants in the field and the laboratory'. In: *Techniques in bioproductivity and photosynthesis*. Ed. by J. Coombes, D. O. Hall, S. P. Long and J. M. O. Scurlock. Oxford: Pergamon Press Ltd. (cit. on p. 131).

Madronich, S. (1993). 'The Atmosphere and UV-B Radiation at Ground Level'. In: *Environmental UV Photobiology*. Ed. by A. R. Young, L. O. Björn, J. Moan and W. Nultsch. Plenum Press, New York. Chap. 1. ISBN: 0-306-44443-7 (cit. on p. 24).

Manney, G. L., M. L. Santee, M. Rex, N. J. Livesey, M. C. Pitts, P. Veefkind, E. R. Nash, I. Wohltmann, R. Lehmann, L. Froidevaux et al. (2011). 'Unprecedented Arctic ozone loss in 2011'. In: *Nature* 478, pp. 469-475. DOI: 10.1038/nature10556 (cit. on p. 10).

Marijnissen, J. P. A. and W. M. Star (1987). 'Quantitative light dosimetry in vitro and in vivo'. In: *Lasers in Medical Science* 2, pp. 235-242. DOI: 10.1007/BF02594166 (cit. on p. 78).

Maritorena, S., A. Morel and B. Gentili (1994). 'Diffuse reflectance of oceanic shallow waters: influence of water depth and bottom albedo'. In: *Limnology and Oceanography* 39.7, pp. 1689-1703. URL: http://www.jstor.org/stable/2838204 (cit. on p. 15).

Markvart, J., E. Rosenqvist, J. M. Aaslyng and C. .-.-O. Ottosen (2010). 'How is Canopy Photosynthesis and Growth of Chrysanthemums Affected by Diffuse and Direct Light?' In: *European Journal of Horticultural Science* 75.6, pp. 253-258. ISSN: 1611-4426 (cit. on pp. 54, 123).

Massonnet, C., D. Vile, J. Fabre, M. A. Hannah, C. Caldana, J. Lisec, G. T. S. Beemster, R. C. Meyer, G. Messerli, J. T. Gronlund et al. (2010). 'Probing the reproducibility of leaf growth and molecular phenotypes: a comparison of three *Arabidopsis* accessions cultivated in ten laboratories'. In: *Plant Physiol* 152, pp. 2142-2157. DOI: 10.1104/pp.109.148338 (cit. on p. 119).

McKinlay, A. F. and B. L. Diffey (1987). 'A reference action spectrum for ultraviolet induced erythema in human skin'. In: *CIE Journal* 6, pp. 17-22 (cit. on pp. 25, 80, 105, 112).

McLeod, A. R. (1997). 'Outdoor supplementation systems for studies of the effects of increased uv-b radiation'. In: *Plant Ecology* 128, pp. 78-92. DOI: 10.1023/A:1009794427697 (cit. on p. 48).

Messenger, D. J., A. R. McLeod and S. C. Fry (2009). 'The role of ultraviolet radiation, photosensitizers, reactive oxygen species and ester groups in mechanisms of methane formation from pectin'. In: *Plant Cell and Environment* 32, pp. 1-9. DOI: 10.1111/j.1365-3040.2008.01892.x (cit. on p. 2).

Millar, D. J. and J. W. Hannay (1986). 'Phytotoxicity of phthalate plasticisers. II. Site and mode of action'. In: *Journal of Experimental Botany* 37, pp. 883-897. DOI: 10.1093/jxb/37.6.898 (cit. on p. 129).

Mitchell, D. L. and D. Karentz (1993). 'The induction and repair of DNA photodamage in the environment'. In: *Environmental UV photobiology*. Ed. by A. R. Young, L. O. Björn, J. Moan and W. Nultsch. New York: Plenum Press, pp. 345-377. ISBN: 0-306-44443-7 (cit. on p. 19).

Mobley, C. D. (1994). *Light and water - radiative transfer in natural waters*. San Diego: Academic Press. URL: http://www.curtismobley.com/lightandwater.zip (cit. on pp. 4, 8, 18, 20, 21).

– (2011). 'Fast light calculations for ocean ecosystem and inverse models'. In: *Optics Express* 19.20, pp. 18927-18944. DOI: 10.1364/OE.19.018927 (cit. on p. 21).

Mobley, C. D. and L. K. Sundman (2003). 'Effects of optically shallow bottoms on upwelling radiances: inhomogeneous and sloping bottoms'. In: *Limnology and Oceanography, Light in Shallow Waters* 48.1, part 2, pp. 329-336. URL: http://www.jstor.org/stable/3597753 (cit. on p. 15).

Mobley, C. D., H. Zhang and K. J. Voss (2003). 'Effects of optically shallow bottoms on upwelling radiances: bidirectional reflectance distribution function effects'. In: *Limnology and Oceanography, Light in Shallow Waters* 48.1, part 2, pp. 337-345. URL: http://www.jstor.org/stable/3597754 (cit. on p. 15).

Möglich, A., X. Yang, R. A. Ayers and K. Moffat (2010). 'Structure and function of plant photoreceptors'. In: *Annu Rev Plant Biol* 61, pp. 21-47. DOI: 10.1146/annurev-arplant-042809-112259 (cit. on p. 1).

Monsi, M. and T. Saeki (1953). 'Über den Lichtfaktor in den Pflanzengesellschaften und seine Bedeutung für die Stoffproduktion'. In: *Japanese Journal of Botany* 14, pp. 22-52 (cit. on p. 11).

Montalti, M., A. Credi, L. Prodi and M. T. Gandolfi (2006). *Handbook of Photochemistry*. 3rd ed. Boca Raton, FL, USA: CRC Press. 664 pp. ISBN: 0824723775 (cit. on p. 54).

Monteith, J. and M. Unsworth (2008). *Principles of Environmental Physics*. 3rd ed. Academic Press. 440 pp. ISBN: 0125051034 (cit. on p. 15).

Morales, L. O., R. Tegelberg, M. Brosché, M. Keinänen, A. Lindfors and P. J. Aphalo (2010). 'Effects of solar UV-A and UV-B radiation on gene expression and phenolic accumulation in Betula pendula leaves'. In: *Tree Physiol* 30, pp. 923-934. DOI: 10.1093/treephys/tpq051 (cit. on p. 55).

Morel, A. (1974). 'Optical properties of pure water and pure sea water'. In: *Optical Aspects of Oceanography*. Ed. by N. G. Jerlov and E. Steemann Nielsen. London: Academic Press, pp. 1-24. ISBN: 0123849500 (cit. on p. 16).

– (1991). 'Light and marine photosynthesis: a spectral model with geochemical and climatological implications'. In: *Progress in Oceanography* 26, pp. 263-306. DOI: 10.1016/0079-6611(91)90004-6 (cit. on pp. 17, 19).

Morel, A. and L. Prieur (1976). 'Analyse spectrale de l'absorption par les substances dissoutes (substances jaunes)'. In: *Publ. CNEXO* 10.Sect. 1.1.11, pp. 1-9 (cit. on p. 16).

– (1977). 'Analysis of variations in ocean colour'. In: *Limnology Oceanography* 22, pp. 709-722. URL: http://www.jstor.org/stable/2835253 (cit. on p. 19).

Morison, J. I. L. and R. M. Gifford (1984). 'Ethylene contamination of CO2 cylinders. Effects on plant growth in CO2 enrichment studies'. In: *Plant Physiology* 75, pp. 275-277. DOI: 10.1104/pp.75.1.275 (cit. on p. 126).

Morris, D. P. and B. R. Hargreaves (1997). 'The role of photochemical degradation of dissolved organic carbon in regulating the UV transparency of three lakes on the Pocono Plateau'. In: *Limnology and Oceanography*

42, pp. 239-249. URL: `http://www.jstor.org/stable/2838552` (cit. on p. 18).

Musil, C. F., L. O. Björn, M. W. J. Scourfield and G. E. Bodeker (2002). 'How substantial are ultraviolet-B supplementation inaccuracies in experimental square-wave delivery systems?' In: *Environmental and Experimental Botany* 47.1, pp. 25-38. DOI: `DOI:10.1016/S0098-8472(01)00108-3` (cit. on p. 48).

Neori, A., M. D. Krom, S. P. Ellner, C. E. Boyd, D. Popper, R. Rabinovitch, P. J. Davison, O. Dvir, D. Zuber, M. Ucko et al. (1996). 'Seaweed biofilter as regulators of water quality in integrated fish-seaweed culture units'. In: *Aquaculture* 141, pp. 183-199. DOI: `10.1016/0044-8486(95)01223-0` (cit. on p. 134).

Nevas, S., A. Teuber, A. Sperling and M. Lindemann (2012). 'Stability of array spectroradiometers and their suitability for absolute calibrations'. In: *Metrologia* 49, S48-S52. DOI: `10.1088/0026-1394/49/2/S48` (cit. on p. 90).

Newsham, K. K., A. R. McLeod, P. D. Greenslade and B. A. Emmett (1996). 'Appropriate controls in outdoor UV-B supplementation experiments'. In: *Global Change Biology* 2, pp. 319-324. DOI: `10.1111/j.1365-2486.1996.tb00083.x` (cit. on pp. 38, 102).

Newsham, K. K., A. R. McLeod, J. D. Roberts, P. D. Greenslade and B. A. Emmet (1997). 'Direct effects of elevated UV-B radiation on the decomposition of Quercus robur leaf litter'. In: *Oikos* 79, pp. 592-602. URL: `http://www.jstor.org/stable/3546903` (cit. on p. 2).

Newsham, K. K., P. Splatt, P. A. Coward, P. D. Greenslade, A. R. McLeod and J. M. Anderson (2001). 'Negligible influence of elevated UV-B radiation on leaf litter quality of *Quercus robur*'. In: *Soil Biology and Biochemistry* 33, pp. 659-665. DOI: `10.1016/S0038-0717(00)00210-8` (cit. on p. 2).

Nobel, P. S. (2009). *Physicochemical and Environmental Plant Physiology*. 4th. Academic Press. 600 pp. ISBN: 0123741432 (cit. on pp. 2, 138).

Nouguier, J., B. Mostajir, E. Le Floc'h and F. Vidussi (2007). 'An automatically operated system for simulating global change temperature and ultraviolet B radiation increases: Application to the study of aquatic ecosystem responses in mesocosm experiments'. In: *Limnology and Oceanography: Methods* 5, pp. 269-279. DOI: `10.4319/lom.2007.5.269` (cit. on p. 136).

Ohde, T. and H. Siegel (2003). 'Derivation of immersion factors for the hyperspectral TriOS radiance sensor'. In: *Journal of Optics A: Pure and Applied Optics* 5.3, pp. L12-L14. DOI: `doi:10.1088/1464-4258/5/3/103` (cit. on p. 94).

Oke, T. R. (1988). *Boundary Layer Climates*. 2nd. Routledge. 464 pp. ISBN: 0415043190 (cit. on p. 138).

Okerblom, P., T. Lahti and H. Smolander (1992). 'Photosynthesis of a Scots Pine Shoot - A Comparison of 2 Models of Shoot Photosynthesis in Direct and Diffuse Radiation Fields'. In: *Tree Physiology* 10.2, pp. 111-125. DOI: `10.1093/treephys/10.2.111` (cit. on p. 54).

Palenik, B., N. M. Price and F. M. M. Morel (1991). 'Potential effects of UV-B on the chemical environment of marine organisms: A review'. In: *Environmental Pollution* 70, pp. 117-130. DOI: `10.1016/0269-7491(91)90084-A` (cit. on p. 19).

Parisi, A., P. Schouten and D. J. Turnbull (2010). 'UV dosimeter based on Polyphenylene Oxide for the measurement of UV exposures to plants and humans over extended periods'. In: *NIWA 2010 UV Workshop: UV Radiation and its Effects - an Update 2010, 7-9 May 2010*. Queenstown, New Zealand (cit. on pp. 78, 79).

Parisi, A., D. J. Turnbull, P. Schouten, N. Downs and T. J. (2010). 'Techniques for solar dosimetry in different environments'. In: *UV radiation in global climate change: measurements, modeling and effects on ecosystems*. Ed. by W. Gao, D. L. Schmoldt and J. R. Slusser. Springer / Shingua University Press, pp. 192-204. ISBN: 978-3-642-03312-4 (cit. on pp. 78, 79).

Parisi, A. V., V. J. Galea and C. Randall (2003). 'Dosimetric measurement of the visible and UV exposures on field grown soybean plants'. In: *Agricultural and Forest Meteorology* 120, pp. 153-160. DOI: `10.1016/j.agrformet.2003.08.012` (cit. on p. 79).

Parisi, A. V. and M. G. Kimlin (2004). 'Personal solar UV exposure measurements employing modified polysulphone with an extended dynamic range'. In: *Photochem Photobiol* 79, pp. 411-415. DOI: `10.1111/j.1751-1097.2004.tb00028.x` (cit. on p. 79).

Parisi, A. V. and J. C. F. Wong (1996). 'Plant canopy shape and the influences on UV exposures to the canopy'. In: *Photochemistry and Photobiology* 63.6, pp. 143-148. DOI: `10.1111/j.1751-1097.1996.tb02434.x` (cit. on p. 11).

Parisi, A. V., J. C. F. Wong and C. Randall (1998). 'Simultaneous assessment of photosynthetically active and ultraviolet solar radiation'. In: *Agricultural and Forest Meteorology* 92, pp. 97-103. DOI: `10.1016/S0168-1923(98)00094-X` (cit. on p. 79).

Parker, C. A. (1953). 'A new sensitive chemical actinometer. 1. Some trials with potassium ferrioxalate'. In: *Proc. Roy. Soc. London* 220A.1140, pp. 104-116. DOI: `10.1098/rspa.1953.0175` (cit. on p. 76).

Passioura, J. (2006). 'The perils of pot experiments'. In: *Functional Plant Biology* 33.12, pp. 1075-1079. DOI: `10.1071/FP06223` (cit. on p. 126).

Paul, N. (2001). 'Plant responses to UV-B: time to look beyond stratospheric ozone depletion?' In: *New Phyto-

logist 150, pp. 5-8. DOI: `10.1046/j.1469-8137.2001.00090.x` (cit. on p. 2).

Paul, N. D., R. J. Jacobson, A. Taylor, J. J. Wargent and J. P. Moore (2005). 'The use of wavelength-selective plastic cladding materials in horticulture: understanding of crop and fungal responses through the assessment of biological spectral weighting functions'. In: *Photochem Photobiol* 81.5, pp. 1052-1060. DOI: `10.1562/2004-12-06-RA-392` (cit. on p. 2).

Pegau, W. S. and J. R. V. Zaneveld (1993). 'Temperature-dependent absorption of water in the red and near-infrared portions of the spectrum'. In: *Limnology and Oceanography* 38 (1), pp. 188-192. URL: `http://www.jstor.org/stable/2837903` (cit. on p. 16).

Petzold, T. (1977). 'Volume scattering functions for selected ocean waters'. In: *Light in the sea*. Ed. by J. Tyler. Dowden, Hutchinson & Ross, Strouddberg, pp. 152-174. ISBN: 0879332654 (cit. on p. 18).

Phoenix, G. K., D. Gwynn-Jones, J. A. Lee and T. V. Callaghan (2003). 'Ecological importance of ambient solar ultraviolet radiation to a sub-arctic heath community'. In: *Plant Ecology* 165, pp. 263-273. DOI: `10.1023/A:1022276831900` (cit. on p. 55).

Pinnel, N. (2007). 'A method for mapping submersed macrophytes in lakes using hyperspectral remote sensing'. PhD thesis. Technische Universität München. URL: `http://mediatum2.ub.tum.de/node?id=604557` (cit. on p. 15).

Poorter, H., J. Bühler, D. van Dusschoten, J. Climent and J. A. Postma (2012). 'Pot size matters: a meta-analysis of the effects of rooting volume on plant growth'. In: *Functional Plant Biology*, pages. DOI: `10.1071/FP12049` (cit. on p. 126).

Poorter, H., F. Fiorani, M. Stitt, U. Schurr, A. Finck, Y. Gibon, B. Usadel, R. Munns, O. K. Atkin, F. Tardieu et al. (2012). 'The art of growing plants for experimental purposes: a practical guide for the plant biologist'. In: *Functional Plant Biology*. DOI: `10.1071/FP12028` (cit. on p. 138).

Poorter, H., K. J. Niklas, P. B. Reich, J. Oleksyn, P. Poot and L. Mommer (2012). 'Biomass allocation to leaves, stems and roots: meta-analyses of interspecific variation and environmental control'. In: *New Phytologist* 193, pp. 30-50. DOI: `10.1111/j.1469-8137.2011.03952.x` (cit. on p. 138).

Pozdnyakov, D. and H. Grassl (2003). *Colour of inland and coastal waters - a methodology for its interpretation*. Berlin/Heidelberg/New York andChichester: Springer Verlag and Praxis Publishing Ltd. (cit. on p. 18).

Prahl, S. A., M. Keijzer, S. L. Jacques and A. J. Welch (1989). 'A Monte Carlo Model of Light Propagation in Tissue'. In: *SPIE Proceedings of Dosimetry of Laser Radiation*

in Medicine and Biology. Ed. by G. J. Müller and D. H. Sliney. Vol. IS 5, pp. 102-111 (cit. on pp. 4, 20).

Prieur, L. and S. Sathyendranath (1981). 'An optical classification of coastal and oceanic waters based on the specific spectral absorption curves of phytoplankton pigments, dissolved organic matter, and other particulate materials'. In: *Limnology and Oceanography* 26.4, pp. 671-689. URL: `http://www.jstor.org/stable/2836033` (cit. on pp. 16, 17).

Quaite, F. E., B. M. Sutherland and J. C. Sutherland (1992). 'Action spectrum for DNA damage in alfalfa lowers predicted impact of ozone depletion'. In: *Nature* 358, pp. 576-578. DOI: `10.1038/358576a0` (cit. on p. 25).

Quan, X. and E. S. Fry (1995). 'Empirical equation for the index of refraction of seawater'. In: *Applied Optics* 34.18, pp. 3477-3480. DOI: `10.1364/AO.34.003477` (cit. on p. 15).

Quinn, G. P. and M. J. Keough (2002). *Experimental Design and Data Analysis for Biologists*. Cambridge, U.K.: Cambridge University Press. xvii + 537. ISBN: 0-521-00976-6 (cit. on pp. 139, 146, 150).

Quintern, L. E., Y. Furusawa, K. Fukutsu and H. Holtschmidt (1997). 'Characterization and application of UV detector spore films: the sensitivity curve of a new detector system provides good similarity to the action spectrum for UV-induced erythema in human skin'. In: *J Photochem Photobiol B* 37, pp. 158-166. DOI: `10.1016/S1011-1344(96)04414-4` (cit. on p. 78).

Quintern, L. E., G. Horneck, U. Eschweiler and H. Bücker (1992). 'A biofilm used as ultraviolet-dosimeter'. In: *Photochemistry and Photobiology* 55, pp. 389-395. DOI: `10.1111/j.1751-1097.1992.tb04252.x` (cit. on p. 78).

Quintern, L. E., M. Puskeppeleit, P. Rainer, S. Weber, S. el Naggar, U. Eschweiler and G. Horneck (1994). 'Continuous dosimetry of the biologically harmful UV-radiation in Antarctica with the biofilm technique'. In: *J Photochem Photobiol B* 22, pp. 59-66. DOI: `10.1016/1011-1344(93)06954-2` (cit. on p. 78).

Rizzini, L., J.-J. Favory, C. Cloix, D. Faggionato, A. O'Hara, E. Kaiserli, R. Baumeister, E. Schäfer, F. Nagy, G. I. Jenkins et al. (2011). 'Perception of UV-B by the *Arabidopsis* UVR8 Protein'. In: *Science* 332.6025, pp. 103-106. DOI: `10.1126/science.1200660` (cit. on p. 1).

Robertson, D. F. (1972). 'Solar ultraviolet radiation in relation to human sunburn and skin cancer'. PhD thesis. University of Queensland (cit. on p. 80).

Robson, T. M., V. A. Pancotto, C. L. Ballaré, O. E. Sala, A. L. Scopel and M. M. Caldwell (2004). 'Reduction of solar UV-B mediates changes in the *Sphagnum capitulum* microenvironment and the peatland microfungal community'. In: *Oecologia* 140, pp. 480–490. DOI: 10.1007/s00442-004-1600-9 (cit. on p. 55).

Roesler, C. S., M. J. Perry and K. L. Carder (1989). 'Modeling in situ phytoplankton absorption from total absorption spectra in productive inland marine waters'. In: *Limnology and Oceanography* 34.8, pp. 1510–1523. URL: http://www.jstor.org/stable/2837036 (cit. on p. 18).

Rousseaux, M. C., R. Julkunen-Tiitto, P. S. Searles, A. L. Scopel, P. J. Aphalo and C. L. Ballaré (2004). 'Solar UV-B radiation affects leaf quality and insect herbivory in the southern beech tree *Nothofagus antarctica*'. In: *Oecologia* 138, pp. 505–512. DOI: 10.1007/s00442-003-1471-5 (cit. on pp. 55, 59).

Rozema, J., J. Vandestaaij, L. O. Björn and M. Caldwell (1997). 'UV-B as an environmental factor in plant life—Stress and regulation'. In: *Trends in Ecology & Evolution* 12, pp. 22–28. DOI: 10.1016/S0169-5347(96)10062-8 (cit. on p. 131).

Ruggaber, A., R. Dlugi and T. Nakajima (1994). 'Modelling radiation quantities and photolysis frequencies in the troposphere'. In: *Journal of Atmospheric Chemistry* 18, pp. 171–210. DOI: 10.1007/BF00696813 (cit. on p. 105).

Rundel, R. D. (1983). 'Action spectra and estimation of biologically effective UV radiation'. In: *Physiologia Plantarum* 58, pp. 360–366. DOI: 10.1111/j.1399-3054.1983.tb04195.x (cit. on pp. 23, 24, 26).

Rupert, C. S. (1974). 'Dosimetric concepts in photobiology'. In: *Photochemistry and Photobiology* 20, pp. 203–212. DOI: 10.1111/j.1751-1097.1974.tb06568.x (cit. on p. 72).

Saitou, T., Y. Tachikawa, H. Kamada, M. Watanabe and H. Harada (1993). 'Action spectrum for light-induced formation of adventitious shoots in hairy roots of horseradish'. In: *Planta* 189, pp. 590–592. DOI: 10.1007/BF00198224 (cit. on p. 45).

Sampath-Wiley, P. and L. S. Jahnke (2011). 'A new filter that accurately mimics the solar UV-B spectrum using standard UV lamps: the photochemical properties, stabilization and use of the urate anion liquid filter'. In: *Plant Cell Environ* 34, pp. 261–269. DOI: 10.1111/j.1365-3040.2010.02240.x (cit. on p. 54).

Sathyendranath, S., L. Prieur and A. Morel (1989). 'A three-component model of ocean colour and its application to remote sensing of phytoplankton pigments in coastal waters'. In: *International Journal of Remote Sensing* 10.8, pp. 1373–1394. DOI: 10.1080/01431168908903974 (cit. on p. 17).

Schindler, D. W., P. J. Curtis, B. R. Parker and M. P. Stainton (1996). 'Consequences of climate warming and lake acidification for UV-B penetration in North American boreal lakes'. In: *Nature* 379, pp. 705–708. DOI: 10.1038/379705a0 (cit. on p. 18).

Schouten, P. W., A. V. Parisi and D. J. Turnbull (2007). 'Evaluation of a high exposure solar UV dosimeter for underwater use'. In: *Photochemistry and Photobiology* 83, pp. 931–937. DOI: 10.1111/j.1751-1097.2007.00085.x (cit. on p. 79).

– (2008). 'Field calibrations of a long-term UV dosimeter for aquatic UV-B exposures'. In: *Journal of Photochemistry and Photobiology, B* 91, pp. 108–116. DOI: 10.1016/j.jphotobiol.2008.02.004 (cit. on p. 79).

– (2010). 'Usage of the polyphenylene oxide dosimeter to measure annual solar erythemal exposures'. In: *Photochemistry and Photobiology* 86, pp. 706–710. DOI: 10.1111/j.1751-1097.2010.00720.x (cit. on p. 79).

Schreiner, M., I. Mewis, S. Huyskens-Keil, M. Jansen, R. Zrenner, J. Winkler, N. O'Brian and A. Krumbein (2012). 'UV-B-induced secondary plant metabolites - potential benefits for plant and human health'. In: *Critical Reviews in Plant Sciences* 31 (3), pp. 229–240. DOI: doi:10.1080/07352689.2012.664979. URL: http://www.tandfonline.com/doi/abs/10.1080/07352689.2012.664979 (cit. on p. 1).

Schwander, H., P. Koepke, A. Ruggaber, T. Nakajima, A. Kaifel and A. Oppenrieder (2000). *System for transfer of atmospheric radiation STAR - version 2000* (cit. on p. 105).

Schwiegerling, J. (2004). *Field guide to visual and ophthalmic optics*. SPIE Press, Bellingham, WA (cit. on p. 6).

Scully, N. M. and D. R. S. Lean (1994). 'The attenuation of ultraviolet radiation in temperate lakes'. In: *Arch. Hydrobiol. Beih.* 43, pp. 135–144 (cit. on p. 18).

Scully, N. M., W. F. Vincent, D. R. S. Lean and W. J. Cooper (1997). 'Implications of ozone depletion for surface-water photochemistry: Sensitivity of clear lakes'. In: *Aquatic Sciences* 59, pp. 260–274. DOI: 10.1007/BF02523277 (cit. on p. 19).

Seckmeyer, G., A. Bais, G. Bernhard, M. Blumthaler, C. R. Booth, P. Disterhoft, P. Eriksen, R. L. McKenzie, M. Miyauchi and C. Roy (2001). *Instruments to Measure Solar Ultraviolet Radiation - Part 1: Spectral Instruments*. Tech. rep. WMO/TD-No. 1066, GAW Report No. 125. Geneva: World Meteorological Organization (cit. on p. 110).

Seckmeyer, G., A. Bais, G. Bernhard, M. Blumthaler, C. R. Booth, K. Lantz, R. L. McKenzie, P. Disterhoft and A. Webb (2005). *Instruments to Measure Solar Ultraviolet Radiation Part 2: Broadband Instruments Measuring Erythemally Weighted Solar Irradiance*. WMO-GAW Report 164. Geneva, Switzerland: World Meteorological Organization (WMO) (cit. on pp. 84, 110).

Seckmeyer, G., A. Bais, G. Bernhard, M. Blumthaler, S. Drüke, P. Kiedron, K. Lantz, R. L. McKenzie, S. Riechelmann, N. Kouremeti et al. (2010). *Instruments to Measure Solar Ultraviolet Radiation - Part 4: Array Spectroradiometers*. GAW Report 191. Geneva: Global Atmosphere Watch, World Meteorological Organization. URL: http://www.wmo.int/pages/prog/arep/gaw/documents/GAW191_TD_No_1538_web.pdf (cit. on pp. 94, 110).

Seckmeyer, G., A. Bais, G. Bernhard, M. Blumthaler, B. Johnsen, K. Lantz and R. McKenzie (2010). *Instruments to Measure Solar Ultraviolet Radiation - Part 3: Multi-channel filter instruments*. Tech. rep. WMO/TD-No. 1537, GAW Report No. 190. Geneva: World Meteorological Organization (cit. on p. 110).

Seckmeyer, G. and H.-D. Payer (1993). 'A new sunlight simulator for ecological research on plants'. In: *Journal of Photochemistry and Photobiology B: Biology* 21.2–3, pp. 175–181. DOI: 10.1016/1011-1344(93)80180-H (cit. on p. 50).

Seliger, H. H. and W. D. McElroy (1965). *Light: Physical and biological action*. New York and London: Academic Press. xi+417. ISBN: 0126358508 (cit. on p. 76).

Setlow, R. B. (1974). 'The wavelengths in sunlight effective in producing skin cancer: a theoretical analysis'. In: *Procceedings of the National Academy of Sciences of the U.S.A.* 71, pp. 3363–3366 (cit. on pp. 25, 78, 111).

Shimazaki, K.-I., M. Doi, S. M. Assmann and T. Kinoshita (2007). 'Light Regulation of Stomatal Movement'. In: *Annual Review of Plant Biology* 58, pp. 219–247. DOI: 10.1146/annurev.arplant.57.032905.105434 (cit. on p. 1).

Shropshire, W. (1972). 'Action spectroscopy'. In: *Phytochrome*. Ed. by K. Mitrakos and W. Shropshire. London: Academic Press, pp. 161–181. ISBN: 0125005504 (cit. on p. 23).

Sliney, D. H. (2007). 'Radiometric quantities and units used in photobiology and photochemistry: recommendations of the Commission Internationale de L'Eclairage (International Commission on Illumination)'. In: *Photochemistry and Photobiology* 83, pp. 425–432. DOI: 10.1562/2006-11-14-RA-1081 (cit. on p. xxiii).

Smith, H. F. (1957). 'Interpretation of adjusted treatment means and regressions in analysis of covariance'. In: *Biometrics* 13, pp. 281–308. URL: http://www.jstor.org/stable/2527917 (cit. on p. 143).

Smith, R. C. and K. S. Baker (1979). 'Penetration of UV-B and biologically effective dose-rates in natural waters'. In: *Photochemistry and Photobiology* 29, pp. 311–323. DOI: 10.1111/j.1751-1097.1979.tb07054.x (cit. on p. 18).

Smith, R. C. and K. S. Baker (1981). 'Optical properties of the clearest natural waters (200-800 nm)'. In: *Applied Optics* 20.2, pp. 177–184. DOI: 10.1364/AO.20.000177 (cit. on p. 16).

Smith, R. C., B. B. Prézelin, K. S. Baker, R. R. Bidigare, N. P. Boucher, T. Coley, D. Karentz, S. MacIntyre, H. A. Matlick, D. Menzies et al. (1992). 'Ozone depletion: Ultraviolet radiation and phytoplankton biology in Antarctic waters'. In: *Science* 255, pp. 952–959. DOI: 10.1126/science.1546292 (cit. on p. 15).

Smith, R. C. and J. E. Tyler (1976). 'Transmission of solar radiation into natural waters'. In: Photochemical and Photobiological Reviews 1. Ed. by K. C. Smith, pp. 117–155 (cit. on p. 16).

Smoluchowski, M. (1908). 'Molekular-kinetische Theorie der Opaleszenz von Gasen im kritischen Zustande, sowie einiger verwandter Erscheinungen'. In: *Annalen der Physik* 25, pp. 205–226 (cit. on p. 16).

Sommaruga, R. and R. Psenner (1997). 'Ultraviolet radiation in a high mountain lake of the Austrian Alps: Air and underwater measurements'. In: *Photochemistry and Photobiology* 65, pp. 957–963. DOI: 10.1111/j.1751-1097.1997.tb07954.x (cit. on p. 18).

Stanghellini, C. (1987). *Transpiration of greenhouse crops—an aid to climate management*. Wageningen, NL: Intituut voor Mechanisatie, Arbeid en Gebouwen (cit. on p. 138).

Stanhill, G. and S. Cohen (2001). 'Global dimming: a review of the evidence for a widespread and significant reduction in global radiation with discussion of its probable causes and possible agricultural consequences'. In: *Agricultural and Forest Meteorology* 107, pp. 255–278. DOI: 10.1016/S0168-1923(00)00241-0 (cit. on p. 10).

Stanhill, G. and H. Z. Enoch, eds. (1999). *Greenhouse Ecosystems, Ecosystems of the world*. Vol. 20. Amsterdam, NL: Elsevier. 434 pp. ISBN: 0444882677 (cit. on p. 138).

Stewart, A. J. and R. G. Wetzel (1980). 'Fluorescence: absorbance ratios–a molecular-weight tracer of dissolved organic matter'. In: *Limnology and Oceanography* 25, pp. 559–564. URL: http://www.jstor.org/stable/2835308 (cit. on p. 18).

Tanskanen, A., A. Lindfors, A. Määttä, N. Krotkov, J. Herman, J. Kaurola, T. Koskela, K. Lakkala, V. Fioletov, G. Bernhard et al. (2007). 'Validation of daily erythemal doses from Ozone Monitoring Instrument with ground-based UV measurement data'. In: *Journal of Geophysical Research (Atmospheres)* 112.D11, D24S44. DOI: 10.1029/2007JD008830 (cit. on p. 99).

Tedetti, M. and R. Sempére (2006). 'Penetration of Ultraviolet Radiation in the Marine Environment: A Review'. In: *Photochemistry and Photobiology* 82, pp. 389–397. DOI: 10.1562/2005-11-09-IR-733 (cit. on p. 95).

Tennessen, D. J., E. L. Singsaas and T. D. Sharkey (1994). 'Light-emitting diodes as a light source for photosynthesis research'. In: *Photosynthesis Research* 39, pp. 85–92. DOI: 10.1007/BF00027146 (cit. on p. 44).

Tevini, M. (1993). 'Effects of Enhanced UV-B Radiation on Terrestrial Plants'. In: *UV-B Radiation and Ozone Depletion: Effects on Humans, Animals, Plants, Microorganisms, and Materials*. Ed. by M. Tevini. Boca Raton: Lewis Publishers, pp. 125–153. ISBN: 0-87371-911-5 (cit. on p. 1).

Thiel, S., T. Döhring, M. Köfferlein, A. Kosak, P. Martin and H. K. Seidlitz (1996). 'A Phytotron for Plant Stress Research: How Far Can Artificial Lighting Compare to Natural Sunlight?' In: *Journal of Plant Physiology* 148.3-4, pp. 456–463. DOI: 10.1016/S0176-1617(96)80279-3 (cit. on p. 50).

Thimijan, R. W., H. R. Carns and L. E. Campbell (1978). *Final Report (EPA-IAG-D6-0168): Radiation sources and related environmental control for biological and climatic effects UV research (BACER)*. Tech. rep. Washington, DC: Environmental Protection Agency (cit. on pp. xxiii, 25, 111).

Tukey, J. W. (1991). 'The Philosophy of Multiple Comparisons'. In: *Statistical Science* 6.1, pp. 100–116. DOI: 10.1214/ss/1177011945 (cit. on p. 139).

Turnbull, D. J. and P. W. Schouten (2008). 'Utilising polyphenylene oxide for high exposure solar UVA dosimetry'. In: *Atmospheric Chemistry and Physics* 8.10, pp. 2759–2762. DOI: 10.5194/acp-8-2759-2008 (cit. on p. 79).

Tyler, J. E. (1968). 'The Secchi disc'. In: *Limnology and Oceanography* 13.1, pp. 1–6. URL: http://www.jstor.org/stable/2833820 (cit. on p. 20).

UNEP (2003). 'Environmental effects of ozone depletion and its interactions with climate change: 2002 Assessment'. In: *Photochemical and Photobiological Sciences* 2, pp. 1–72 (cit. on p. 27).

– (2007). 'Environmental effects of ozone depletion and its interactions with climate change: 2006 Assessment'. In: *Photochemical and Photobiological Sciences* 6.3, pp. 201–332 (cit. on p. 27).

– (2011). *2010 assessment report of the Environmental effects of ozone depletion and its interactions with climate change*. Photochemical and Photobiological Sciences 10(2), 165–320. Also published by UNEP (cit. on pp. 1, 27).

Urban, O., D. Janous, M. Acosta, R. Czerny, I. Markova, M. Navratil, M. Pavelka, R. Pokorny, M. Sprtova, R. Zhang et al. (2007). 'Ecophysiological controls over the net ecosystem exchange of mountain spruce stand. Comparison of the response in direct vs. diffuse solar radiation'. In: *Global Change Biology* 13, pp. 157–168. DOI: 10.1111/j.1365-2486.2006.01265.x (cit. on p. 54).

Urban, O., K. Klem, A. Ac, K. Havránková, P. Holisová, M. Navrátil, M. Zitová, K. Kozlová, R. Pokorný, M. Sprtová et al. (2012). 'Impact of clear and cloudy sky conditions on the vertical distribution of photosynthetic CO_2 uptake within a spruce canopy'. In: *Functional Ecology* 26, pp. 46–55. DOI: 10.1111/j.1365-2435.2011.01934.x (cit. on p. 54).

Van den Boogaard, R., J. Harbinson, M. Mensink and J. Ruijsch (2001). 'Effects of quality and daily distribution of irradiance on photosynthetic electron transport and CO2 fixation in tomato'. In: *Proceedings of the 12th International Congress on Photosynthesis, Brisbane, Australia*. Vol. S28-030, pages (cit. on p. 124).

Veit, M., T. Bilger, T. Muhlbauer, W. Brummet and K. Winter (1996). 'Diurnal changes in flavonoids'. In: *Journal of Plant Physiology* 148.3-4, pp. 478–482. DOI: 10.1016/S0176-1617(96)80282-3 (cit. on p. 21).

Venables, W. N. and B. D. Ripley (2000). *S Programming*. Statistics and Computing. New York: Springer. x + 264. ISBN: 0 387 98966 8 (cit. on p. 116).

Venables, W. N. and B. D. Ripley (2002). *Modern Applied Statistics with S*. 4th. New York: Springer. 512 pp. ISBN: 0-387-95457-0 (cit. on p. 116).

Villafañe, V. E., K. Sundbäck, F. L. Figueroa and E. W. Helbling (2003). 'Photosynthesis in the aquatic environment as affected by UVR'. In: *UV effects in aquatic organisms and ecosystems*. Ed. by E. W. Helbling and H. E. Zagarese. Cambridge: The Royal Society of Chemistry, pp. 357–397. ISBN: 0-85404-301-2 (cit. on p. 131).

Vincent, W. F. and S. Roy (1993). 'Solar ultraviolet-B radiation and aquatic primary production: Damage, protection, and recovery'. In: *Environmental Reviews* 1, pp. 1–12. DOI: 10.1139/a93-001 (cit. on p. 19).

Visser, A. J., M. Tosserams, M. W. Groen, G. W. H. Magendans and J. Rozema (1997). 'The combined effects of CO_2 concentration and solar UV-B radiation on faba bean grown in open-top chambers'. In: *Plant, Cell and Environment* 20.2, pp. 189–199. DOI: 10.1046/j.1365-3040.1997.d01-64.x (cit. on p. 127).

Vogelmann, T. C. and L. O. Björn (1984). 'Measurement of light gradients and spectral regime in plant tissue with a fiber optic probe'. In: *Physiologia Plantarum* 60, pp. 361–368. DOI: 10.1111/j.1399-3054.1984.tb06076.x (cit. on p. 21).

Vogelmann, T. C. and J. R. Evans (2002). 'Profiles of light absorption and chlorophyll within spinach leaves from chlorophyll fluorescence'. In: *Plant Cell and Environment* 25, pp. 1313–1323. DOI: 10.1046/j.1365-3040.2002.00910.x (cit. on p. 21).

Vogelmann, T. C. and T. Han (2000). 'Measurements of gradients of absorbed light in spinach leaves from chlorophyll fluorescence profiles'. In: *Plant Cell and*

Environment 23, pp. 1303-1311. DOI: 10.1046/j.1365-3040.2000.00649.x (cit. on p. 21).

Wang, L.-. H., S. L. Jacques and L.-. Q. Zheng (1995). 'MCML - Monte Carlo modeling of photon transport in multilayered tissues'. In: *Computer Methods and Programs in Biomedicine* 47, pp. 131-146. DOI: 10.1016/0169-2607(95)01640-F (cit. on pp. 4, 21).

Wargent, J. J., V. C. Gegas, G. I. Jenkins, J. H. Doonan and N. D. Paul (2009). 'UVR8 in *Arabidopsis thaliana* regulates multiple aspects of cellular differentiation during leaf development in response to ultraviolet B radiation'. In: *New Phytologist* 183.2, pp. 315-326. DOI: 10.1111/j.1469-8137.2009.02855.x (cit. on p. 1).

Watanabe, M., M. Furuya, Y. Miyoshi, Y. Inoue, I. Iwahashi and K. Matsumoto (1982). 'Design and Performance of The Okazaki Large Spectrograph for Photobiological Research'. In: *Photochemistry and Photobiology* 36, pp. 491-498. DOI: 10.1111/j.1751-1097.1982.tb04407.x (cit. on p. 45).

Webb, A., J. Gröbner and M. Blumthaler (2006). *A Practical Guide to Operating Broadband Instruments Measuring Erythemally Weighted Irradiance.* Tech. rep. Produced by the joint efforts of WMO SAG UV, Working Group 4 of COST-726 Action "Long Term Changes and Climatology of UV Radiation over Europe" (cit. on p. 81).

Webb, A. R., H. Slaper, P. Koepke and A. W. Schmalwieser (2011). 'Know your standard: clarifying the CIE erythema action spectrum'. In: *Photochemistry and Photobiology* 87, pp. 483-486. DOI: 10.1111/j.1751-1097.2010.00871.x (cit. on pp. 80, 112).

Wehrli, C. (1985). *Extraterrestrial solar spectrum.* PMOD-/WRC Publication 615. Physikalisch-Meteorologisches Observatorium und World Radiation Center Davos Dorf, Switzerland (cit. on p. 10).

WHO (2002). *Global solar UV index: a practical guide.* Tech. rep. ISBN 92 4 159007 6. World Health Organization. URL: http://www.unep.org/PDF/Solar_Index_Guide.pdf (cit. on p. 35).

WMO (2008). *Guide to Meteorological Instruments and Methods of Observation, WMO-No. 8.* Tech. rep. Seventh edition. World Meteorological Organization (cit. on p. 82).

Wozniak, B. and J. Dera (2007). *Light absorption in sea water.* Vol. 33. Atmospheric and Oceanographic Sciences Library. Springer, Dordrecht. 452 pp. ISBN: 0387307532 (cit. on pp. 15, 19).

Wu, D., Q. Hu, Z. Yan, W. Chen, C. Yan, X. Huang, J. Zhang, P. Yang, H. Deng, J. Wang et al. (2012). 'Structural basis of ultraviolet-B perception by UVR8'. In: *Nature* 484, 214--219. DOI: 10.1038/nature10931 (cit. on p. 1).

Wu, M., E. Grahn, L. A. Eriksson and A. Strid (2011). 'Computational evidence for the role of *Arabidopsis thaliana* UVR8 as UV-B photoreceptor and identification of its chromophore amino acids'. In: *Journal of Chemical Information and Modeling* 51, pp. 1287-1295. DOI: 10.1021/ci200017f (cit. on p. 1).

Yan, N. D., W. Keller, N. M. Scully, D. R. S. Lean and P. J. Dillon (1996). 'Increased UV-B penetration in a lake owing to drought-induced acidification'. In: *Nature* 381, pp. 141-143. DOI: 10.1038/381141a0 (cit. on p. 18).

Ylianttila, L., R. Visuri, L. Huurto and K. Jokela (2005). 'Evaluation of a single-monochromator diode array spectroradiometer for sunbed UV-radiation measurements'. In: *Photochemistry and Photobiology* 81, pp. 333-341. DOI: 10.1562/2004-06-02-RA-184 (cit. on p. 94).

Zepp, R. G. (1982). 'Photochemical transformations induced by solar ultraviolet radiation in marine ecosystems'. In: *The role of solar ultraviolet radiation in marine ecosystems.* Ed. by J. Calkins. New York: Plenum Press, pp. 293-307 (cit. on p. 19).

Zuur, A. F., E. N. Ieno and E. Meesters (2009). *A Beginner's Guide to R.* Springer. 236 pp. ISBN: 0387938362 (cit. on p. 150).

Zuur, A. F., E. N. Ieno, N. Walker, A. A. Saveliev and G. M. Smith (2009). *Mixed Effects Models and Extensions in Ecology with R.* New York: Springer. 596 pp. ISBN: paperback 1441927646, hardcover 978-0-387-87457-9 (cit. on p. 150).

Glossary

absorbance $A = \log E_0/E_1$, where E_0 is the incident irradiance, and E_1 is the transmitted irradiance. xxiii, 16, 21, 52, 77, 78

absorptance radiation that is absorbed by an object, as a fraction of the incident irradiance: $\alpha = E_{abs}/E_0$, where E_0 is the incident irradiance and E_{abs} is the absorbed irradiance. xxiii, 52, 65, 66

biological spectral weighting function a function used to estimate the biological effect of radiation. It is convoluted—i.e. multiplied wavelength by wavelength—with the spectral irradiance of a source of UV radiation to obtain a biologically effective irradiance. xxiii, 27, 80, 83, 100, 101, 105, 111, 113

biologically effective exposure radiation exposure (also called dose by biologists) measured according to the effectiveness of radiation in producing a certain biological response. 73

collimated radiation is collimated when it is emitted by a point source, e.g. the sun or a star, when observed from far away. Also a slide projector produces a bean of light that is rather well collimated. The opposite is diffuse radiation, when it arrives from many directions, e.g. the radiation under an overcast sky. 20, 45, 71, 72, 77

direct radiation solar radiation that arrives directly at the ground level, without being scattered by gases and particles of the atmosphere. 9, 11, 55, 71, 97, 134

global radiation total solar radiation arriving at ground level. It is the sum of direct and diffuse radiation. 9, 97, 99, 123

isotropic radiation is isotropic when it arrives equally from all directions, e.g. it is completely diffuse. 72

monochromator an optical device that spreads the incoming optical radiation according to its wavelength. An example of a monochromator is a prism. Most spectroradiometers use ruled or holographic gratings as a monochromator. 44

photosynthetic photon flux density another name for 'PAR photon irradiance'. xxiv, 72, 113, 123

photosynthetically active radiation radiation driving photosynthesis in higher plants, it describes a wavelength range—i.e. $\lambda = 400-700$ nm—but does not define whether an energy or photon quantity is being used. xxiv

proportional-integral-derivative a *proportional integral derivative* controller (PID controller) is a control loop feedback mechanism. A PID controller calculates an "error" value as the difference between a measured process variable and a desired setpoint. The controller attempts to minimize the error by adjusting the process control inputs. A well tuned PID controller (with correct parameters) minimizes overshoot and transient deviations, by adjusting, for example, the dimming in a modulated system based on the size of the error and the response characteristics of the controlled system. xxiv, 48

radiation amplification factor gives the percent change in biologically effective UV irradiance for a 1% change in stratospheric ozone column thickness. Its value varies with the BSWF used in the calculation. xxiv, 95, 101

reflectance radiation that is reflected by an object, as a fraction of the incident irradiance: $\rho = E_{rfl}/E_0$, where E_0 is the incident irradiance and E_{rfl} is the reflected irradiance. xxiii, 52

scattered or 'diffuse' radiation solar radiation that arrives at ground level after being scattered by gases and particles of the atmosphere, also called 'diffuse radiation'. 9, 55, 71, 134

spectrometer an instrument for measuring spectra of radiation. Depending on the set-up some spectrometers can be used as spectroradiometers or as spectrophotometers. 85

spectrophotometer a spectrometer equiped with a light source and used to measure absorbance, absorptance, transmittance and reflectance of objects, e.g. the absorbance spectrum measured from a solution contained in a cuvette can be used to estimate the concentrations of solutes that absorb radiation. 85

spectroradiometer a spectrometer equiped with an entrance optics suitable for measuring radiation, e.g. with a cosine diffuser it can be used to measure spectral irradiance. 85

stray light the unwanted radiation of a different wavelength than that which is being measured, that reaches the sensor of a spectrometer. It usually originates from internal reflections within the instrument. A good single-monochromator spectrometer may have stray light of about 0.1% of the measured signal under ideal conditions, while a double-monochromator spectrometer may have a stray light of only about 0.0001% of the measured signal. 44

transmittance radiation that is transmitted by an object, as a fraction of the incident irradiance: $\tau = E_{trs}/E_0$, where E_0 is the incident irradiance and E_{trs} is the transmitted irradiance. xxiii, 52, 65, 66

Index